India and China

An Advanced Technology Race and How the United States Should Respond

by

Ernest H. Preeg

Library of Congress Cataloging-in-Publication Data

Preeg, Ernest H.
India and China : an advanced technology race and how the United States should respond / by Ernest H. Preeg ; foreword by Thomas J. Duesterberg and John J. Hamre.
p. cm.
Includes bibliographical references and index.
ISBN 978-0-9745674-3-3 (pbk. : alk. paper)
1. High technology industries–India. 2. High technology industries–China. 3. Technological innovations–Government policy–United States. 4. India–Economic policy. 5. China–Economic policy. 6. United States–Economic policy. I. Title: India and China.
HC440.H53P74 2008
338'.0640973—dc22 2008003378

Cover design by Rensford L. Lovell

Library of Congress Catalog Card Number: 2008003378
ISBN: 978-0-9745674-3-3

First published in 2008

Manufacturers Alliance/MAPI
1600 Wilson Boulevard, Suite 1100
Arlington, VA 22209-2594
www.mapi.net

Printed in the United States of America
Signature Book Printing, www.sbpbooks.com

For Caleb
and
Future American Leaders
In Our Brave New
Advanced Technology World

CONTENTS

ILLUSTRATIONS

FIGURES

TABLES

FOREWORD

by
Thomas J. Duesterberg and John J. Hamre

In the 15th century, China ruled the Eastern seas and India was a major trader and entrepot in the important East-West trading routes. That era of economic history is only dimly understood and largely forgotten. The balance of economic forces in that distant era, however, inasmuch as we can reconstruct it, is weighted strongly in favor of the two Asian giants. According to Maddison, in 1500 China and India together accounted for nearly 50 percent of total world output.[i] Europe, by contrast, had about 20 percent of global product, and North America was just a rounding error in the global economy.

Despite the gradual colonization of much of Asia, dominance of global trade by European merchants in the next three centuries, and the appearance of the Americas on the world scene, by the dawn of the industrial revolution in 1800 the weight of the two Asian super giants in the global economy was still estimated to be nearly 50 percent in 1820. And this despite the mysterious and sudden withdrawal of China from world trade in the late 15th century. While this heavy economic presence is largely accounted for by the huge demographic dominance of the Asian powers, even an acute reporter such as the great French historian Fernand Braudel, writing in the late 20th century, argued that ***per capita*** income in China and India was at least equal to that of the Euro-centric world as late as the Napoleonic era.[ii]

> For it can hardly be questioned that until the nineteenth century, the rest of the world outweighed Europe both in population . . . and, in wealth, if it is virtually beyond question that Europe was less rich than the worlds it was exploiting . . . we still do not really know how this position of superiority was established and above all maintained—for the gap grew steadily.

Braudel put per capita income in China and India at a little over $200 around 1800 (in 1960 U.S. dollars), while that of France around the time of the Revolution was $170-$200 and England in 1700 was $150

---

[i] Angus Maddison, *The World Economy: A Millennial Perspective (Paris, OECD, 2001), p. 263.*

[ii] Fernand Braudel, *The Perspective of the World* (New York, Harper & Row, 1984), pp. 533-534.

to $190. Maddison, a more careful statistician with access to much subsequent research, has Europe surpassing Asia in per capita GDP by 1500.[iii]

For our purposes, it matters less who is right in this historical dispute. There is, however, a clear link between economic might and strategic influence in the world. What is still surprising is the realization that the global giants of the 15th century had lost their dominance a few short centuries later. For by the time they reached the nadir of their economic power in 1950, China and India together represented only about 9 percent of global output, outpaced by the 30 percent of the United States and 26 percent in Western Europe. China lost its relative supremacy in large part by withdrawing from the world, and India lost out to the Europeans in the technology and entrepreneurial race to dominate world trade. Their dominance was never dependant on global trade, but their withdrawal or under-performance in this arena goes far in explaining their relative decline, both in economics and in global influence.

It is only in the last two generations that these "sleeping giants," to paraphrase Napoleon on China, have fully awakened and begun to reassert their tremendous size and breadth in the world economy and world affairs. By the close of the millennium, they had greatly narrowed the gap, although by how much depends on the basis of measurement—either in terms of exchange rates or purchasing power parity, as addressed in the annex to Chapter 1 of Dr. Preeg's book.[iv] One of the primary strategies for achieving such a remarkable recovery, which has only accelerated in the new century, was by comprehensively reengaging in the global economy. And their success is inextricably and irreversibly linked to their growing participation in what has become a global economy. Their newfound economic success is a foundation for growing geostrategic influence. Unlike five centuries ago, a withdrawal would lead to absolute and relative decline in their standard of living—and probably to large-scale political instability.

This book is a cogent summary of the story of this reengagement, and an exploration of the implications of the reemergence of the Asian giants to the "pax Americana" which emerged—both in terms of a political order and an economic order—in the 20th century. Much of the 21st century will be engaged in the effort on all sides to build a new economic and geopolitical architecture to accommodate the rise of India and China. The story of China's rise is relatively well known and

---

[iii] Maddison, *The World Economy*, *op. cit.*, p. 264.

[iv] Such different measurements are not available, or at least are much more crude, for periods before the 19th century.

has been told by the present author, among many others.[v] The story of India, however, is less well known and deserves new emphasis, especially, as Dr. Preeg contends, if it is more likely to sustain its high rate of growth than will China in future years.

The first half of the book chronicles and analyzes the economic development of China since 1978 and of India since 1991. Both countries have been growing at a remarkable 8 to 11 percent per year and taking an ever larger place on the stage of the global economy. The author places this rise in the context—of which it is of course a major part—of the technology-driven transformation of the global economy. What is perhaps not as well understood, as Dr. Preeg convincingly demonstrates it should be, is the speed at which China and India are ascending the technology ladder to become world class competitors in both advanced manufacturing and business services. These emerging giants are not just demographic powerhouses exploiting their abundant supplies of cheaper labor. They are skillfully exploiting, too, their growing and historically well-grounded educational systems to become competitive in a wide array of advanced technology industries and services. So, not only are they gaining global market share, but they are doing this in advanced technology sectors such as information technology, aerospace, and engineering services. For this reason alone, the United States, Europe, and other industrialized nations need to pay attention and respond in a constructive way that contributes to strengthening the global trading and financial systems.

In the second half of the book, Dr. Preeg explores a number of responses to this historic turning point: in terms of domestic policy in the United States, in terms of the institutions of global cooperation and development such as those originating in the Bretton Woods agreements of 1944, and in terms of global trade and national security arrangements.

The U.S. economic relationships with China and India offer large mutual benefits from increased trade and investment, although current problems exist, particularly with China. U.S. export competitiveness in manufactures and related services and long-standing leadership in technological innovation are at risk due especially to the rise of the Chinese manufacturing juggernaut. India is increasingly a challenger for engineering, software, and some other business services, although Dr. Preeg chronicles their efforts to become competitive in industry as well. Dr. Preeg argues that the United States needs to respond more forcefully, in terms of currency misalignment and a number of trade policy issues on the international front, and through domestic economic

---

[v] Ernest Preeg, *The Emerging Chinese Advanced Technology Superstate* (Manufacturers Alliance/MAPI and Hudson Institute, 2005).

policies that strengthen U.S. international competitiveness. He explores a number of these strategies for renewal, including for education, public funding for basic research, corporate tax reform, and tort reform. He lays out a comprehensive U.S. response in each of these areas that is calibrated to the emerging global competitive environment in which China and India are destined to play such a major role.

The post-war structure of the international financial and trading systems is also at risk due to the disruption flowing from the entry by the new superstates into the global system, and U.S. relations with China and India will be decisive for the eventual outcome. A broad set of proposals for revising the International Monetary Fund (IMF) and the World Trade Organization (WTO) is presented in the study which respond to the new economic landscape. In terms of international trade arrangements, the continued U.S. pursuit of free trade agreements across the Pacific, including with India, as building blocks toward a multilateral free trade agreement, at least for nonagricultural trade, should be a high priority U.S. trade objective, according to the author. Thus, he counsels deeper engagement on the part of the United States (and Europe and Japan as well) instead of confrontation or withdrawal as part of a constructive response to the vastly changed 21st century economic environment.

In terms of national security and military modernization, U.S. relationships within the "New Asia-Pacific Triangle" are not well-defined, beyond the adversarial U.S.-China relationship over Taiwan. Rapid military modernization in both China and, more recently, India, including deepening integration between defense and advanced technology civilian industries, needs higher priority attention. Such modernization is especially noteworthy for the Chinese and Indian navies, in view of modern civilian shipyards in place or under construction, and important relationships will develop among the three "blue water fleets" over the next several years, across the Pacific and into the Bay of Bengal. For the first time since the 15th century, China, especially, but increasingly India as well, are preparing to assume their historical roles as naval powers. This is already seen in the China and Indian Oceans, but on the not-too-distant horizon one can see these powers in the Pacific, if not the Atlantic. Only dimly understood is the role of information technology in the warfare of the future, but Dr. Preeg also points to the growing competitiveness of China, and to a lesser extent India, in this arena as well. In this arena, too, the emergence of the new technology superpowers is a national security challenge.

We are entering into a truly "new world order," and over the next 10 to 20 years center stage will increasingly be occupied by the United

States, China, and India, the three advanced technology superstates in economic terms, as well as the European Union, an economic superpower but not a superstate. In political terms, it currently involves the two largest democracies and the largest authoritarian state, and the course ahead for democratization within China will greatly influence the ability of the three superstates to provide international leadership. Dr. Preeg provides some beacons of light for the direction of the U.S. response, while suggesting a constructive, and engaged, role for China and India as well.

This study presents all of these issues in a comprehensive and forward-looking manner. It is a clarion call for greater public attention at the outset of an election year for a new Administration with the requirement for new foreign policy formulation less than a year away.

PREFACE

The biggest challenge for this two-year study has simply been to keep up with the rapid pace of events as India and China move up the ladder to become advanced technology superstates. In India, for example, a new law of February 2006 greatly liberalized the establishment of special economic zones (SEZs), and within six months 400 applications for and over 200 approvals of such largely free trade zones had sprung forth. Meanwhile, foreign direct investment (FDI), long dormant, surged to $8 billion in 2005 and to $19 billion in 2006. These and other recent developments led to the unanticipated conclusion that India is highly likely to sustain an 8 to 10 percent annual rate of growth.

In China, in contrast, threatening clouds have arisen about the course ahead. The current account surplus has veered out of control—from 4 percent of gross domestic product in 2004 to 7 percent in 2005, 9 percent in 2006, and a projected 12 percent in 2007—and it will now have to decline. The 2006 revised economic strategy of "independent innovation," essentially favoring Chinese over foreign invested firms, raises further questions about growth-generating export performance, dominated by foreign firms. These and other recent developments have led to the conclusion that China is headed for a "hard landing" adjustment away from excessively export-oriented growth, including at least a couple of years of substantially lower growth.

These country lines of assessment converge on the subtitle theme of an advanced technology race, with China well in the lead in most respects, but with India projected to narrow the gap over the short and longer terms. The original working title was "The Indian Tortoise and the Chinese Hare," with some interesting parallels with the fabled race, but yet again the recent pace of change has overtaken Aesop. India has now shed its tortoise shell and become more fleet afoot.

The other half of the subtitle, "How the United States Should Respond," is likewise a moving target, and the comprehensive set of policy conclusions and recommendations presented here should be taken as a current fix, to be adjusted and extended as the forces in play continue to evolve. There should be no question, however, that important U.S. commercial, national security, and foreign policy interests are at stake, especially for the deeply troubled international financial and trading systems, with China and India heavily if not decisively engaged. Longer term, moreover, the emerging "New Asia-Pacific Triangle" will certainly rise to center stage within truly new international political and economic orders.

This study benefited greatly from the joint sponsorship by the Center for Strategic and International Studies (CSIS) and the Manufacturers Alliance/MAPI. CSIS President John Hamre and my former CSIS colleague Robin Niblett were highly supportive at the organizational stage. CSIS also has in-depth expertise for both India and China. Ambassador Teresita Schaffer, Director of the South Asia Program, and Freeman China Chairholder Bates Gill were invaluable in providing me source material, political counseling, and insightful comments on early drafts of the text. Tezi was especially helpful in preparing my trip to New Delhi, Bangalore, and Mumbai, for setting priority objectives and facilitating a wide range of high quality meetings. The final product also benefits from the broader policy perspectives provided by John Hamre, as co-author of the Foreword, and in the commentary by Tezi.

My professional colleagues at MAPI, as always, provided support and guidance, with a focus on U.S. business interests and engagement in India and China, highlighting policy issues that need to be addressed. I received unstinting support from research assistants Jessica Rushing and Brett Kelly in tracking down source material and translating wide ranging quantitative material into tabular form. Jessica's recent work experience in India on science and technology policy was particularly helpful. Jane Dove provided the indispensable administrative and secretarial support for keeping the project on an orderly, productive track. Most of all, I thank MAPI President Tom Duesterberg, an esteemed colleague of many years, for his encouragement and support throughout, including detailed discussion of and comments on all parts of the text, and his co-authorship of the Foreword.

I also owe special thanks to Indian private sector organizations, the Confederation of Indian Industry (CII), the Federation of Indian Chambers of Commerce and Industry (FICCI), and the National Association of Software and Service Companies (NASSCOM), for detailed briefings and copious background materials on the rapidly modernizing Indian private sector. The CII, in particular, arranged a broad set of meetings with private sector leaders during my India trip, together with logistical guidance and support to keep me on a fully booked schedule without mishap.

Lastly, I thank the Sloan Foundation for financial support for this project from the Sloan Industry Studies Program, and especially Gail Pesyna and Patrick Mulloy for their dedicated interest in issues of greatest consequence for maintaining U.S. export competitiveness and leadership in technological innovation.

CHAPTER 1

Introduction and Principal Conclusions

China and India have both been achieving extraordinarily high economic growth of 8 to 11 percent per year over the past several years, driven by the rapid development of advanced technology industries open to international trade and investment. China got off to a much earlier start in its market-oriented reform program and is far ahead of India in terms of gross domestic product (GDP) and trade. India, however, has recently accelerated the pace of economic reform, with a strong private sector response, and is moving to narrow the Chinese lead. The two nations, moreover, are entering into an advanced technology race, competing for foreign direct investment and export markets for manufactures and business services.

This dynamic and increasingly interactive growth performance on the part of the two largest nations, which together account for a third of the global population, raises a number of far-reaching questions. Can this 8 to 11 percent growth path be sustained and, if so, what challenges will have to be met? Will Chinese and Indian firms rise to become competitive multinational corporations, with leading-edge technological innovation programs? What will the impact be on U.S. commercial, national security, and foreign policy interests? How should the United States respond to strengthen its own export competitiveness and maintain its long-standing leadership in technological innovation? In broader foreign policy terms, how should the United States engage in this advanced technology race between democratic India and communist China?

This study provides answers to these questions and is in two parts. The first part, Chapters 2 through 6, is analytic, and examines recent developments in China and India and the prospect ahead, in fairly specific terms over the coming 2 to 5 years, and more broadly over a longer 10 to 20 year time horizon. The second part, Chapters 7 through 11, presents a comprehensive set of recommendations for the U.S. policy response to the momentous developments under way within China and India and their far-reaching impact on the global economic and political orders.

The study is a sequel to the 2005 work, *The Emerging Chinese Advanced Technology Superstate*,[1] with such superstate status defined

[1] Ernest H. Preeg, *The Emerging Chinese Advanced Technology Superstate* (Manufacturers Alliance/MAPI and Hudson Institute, 2005).

as a nation that will almost certainly achieve great economic, technological, and financial status, that will very likely become financially and politically powerful in international affairs, and that will inevitably strive to become a military superpower as well.[2] That work concluded that China had already crossed the Rubicon toward becoming an advanced technology superstate.

This present work updates the outlook for China and includes India as a second prospective advanced technology superstate. Comparative performance and deepening interactions between the two are highlighted throughout. Much, in fact, has happened in China as well as in India since 2005, and the assessment here goes considerably beyond that contained in the earlier work. The Chinese economy has careened out of balance, with its growth excessively export-oriented, and it faces a "hard landing" to more balanced growth over the next several years. Moreover, China's revised economic strategy of "indigenous" innovation and exports, favoring Chinese over foreign-invested firms, is risky for maintaining high levels of foreign investment in advanced technology industries. India, meanwhile, has experienced more balanced growth, with step-by-step economic reforms creating a more favorable climate for private investment. Foreign direct investment in India, in particular, is surging in many sectors, including manufacturing, to supplement burgeoning indigenous private investment. Both nations have a rising sense of national purpose and destiny to become advanced technology superstates.

The U.S. policy response to these new economic circumstances, within what is termed the new Asia-Pacific triangle, needs to be comprehensive and forceful, involving international financial, trade, and investment policies, and a corresponding domestic policy agenda. Long-standing U.S. technological leadership is at risk, and the United States is losing export competitiveness in manufactures and related business services, the sectors from which most technological innovation derives. In October 2005, a select committee of the National Academies of Science, with ample references to China and India, concluded:[3]

> Having reviewed trends in the United States and abroad, the Committee is deeply concerned that the scientific and technical building blocks of our economic

---

[2] This definition is from Herman Kahn, *The Emerging Japanese Superstate: Challenge and Response* (Prentice-Hall, 1970), p. vii.

[3] *Rising Above the Gathering Storm: Energizing and Employing America for a Brighter Economic Future*, National Academy of Sciences, Preliminary Report, October 2005.

> leadership are eroding at a time when many other nations are gathering strength . . . we fear the abruptness with which a lead in science and technology can be lost—and the difficulty of recovering a lead once lost, if indeed it can be regained at all.

The present study addresses a wide range of specific economic policy issues, both international and domestic, and offers proposals for immediate action. There are also longer term relationships that need to be addressed in a global economy that is undergoing an historic, technology-driven transformation, with China and India playing central roles. The policy challenges include the need for a basic restructuring of the international financial and trading systems, which are currently in functional disarray. In broadest terms, the global political and economic orders are increasingly oriented toward the four advanced technology regions of North America, West Europe, East Asia, and South Asia, with an ever more powerful economic hegemon within each region: the United States, the EU, China, and India. The future course of relationships among these four regions, and among the four advanced technology hegemons in particular, is the principal subject of the concluding chapter.

* * *

This introductory chapter begins with a brief account of the launching of the market-oriented reform programs in China in 1979 and in India in 1991, which are important for understanding the later course of events in both countries up to the present. This is followed by a summary of the principal analytic findings and policy conclusions and recommendations. The chapter concludes with an annex on the critical measurement decision for understanding the relative economic strengths and performance of the United States, China, and India, on the purchasing power parity (PPP) versus the exchange rate measure. This measurement issue has often been misunderstood or wrongly applied, and the debate about which measure to use now benefits from the greatly improved PPP estimates in the December 2007 report of the International Comparison Program.

Launching the Economic Reforms

The launches of radical economic reform programs in China and India, away from state control and protectionism toward market-oriented economies open to international trade and investment, are fascinating stories in themselves. They also involved several parallels, in political and economic terms, which together were decisive in shaping the outcome. Moreover, the same parallel forces, although in altered form, continue to interact in shaping the course of reform in both countries. It is therefore fitting to begin this study with a brief account of how the reforms were launched and a commentary on the parallel forces in play, then and now, which set the stage for the later discussion of industrial modernization in the two countries.

The Chinese Launch

Mao Zedong died in September 1976, the "Gang of Four" was arrested in October, and Deng Xiaoping quickly consolidated his power base and launched a program of market-oriented economic reforms that has been described as "an astonishingly rapid decline in the revolutionary commitment in Chinese society."[4] The precipitating causes for the reforms were both political and economic. China had suffered greatly from the ill-conceived Great Leap Forward in 1958, when 20 million to 30 million people starved to death, and from the purges of the Cultural Revolution of 1966-1976, which decimated the professional and managerial classes. Industrial stagnation resulted and a large trade deficit in 1976 called for immediate economic action, but the official response of a Ten Year Plan, based on huge public investment with no available financing, had no credibility. This financial impasse, moreover, was exaggerated by Deng and his fellow reformers in order to gain greater public support for radical change. They also made invidious comparison with market-oriented high growth in neighboring economies—the Four Tigers of Hong Kong, Singapore, South Korea, and Taiwan.

Deng's reform program was couched in terms of the "Four Modernizations," first presented by Chou Enlai in 1975 and enshrined by Deng at the decisive Party Central Committee meeting in December 1978. They were, in priority order: agriculture, industry, national defense, and science and technology. In agriculture, communes and collective farms were dismantled and replaced with family farms. Limits for private plots were increased and rural markets flourished. Wage labor was introduced and large farms, with hired workers, became prominent. For industry, initial reforms were more limited in scope. Decision making, including labor relations, was delegated to some state enterprises, and they were expected to show a profit. Modernizations three and four—national defense and science and

---

[4] Maurice Meisner, *The Deng Xiaoping Era: An Inquiry Into the Fate of Chinese Socialism 1978-1994* (Hill and Wang, 1996), p. 102.

technology—were essentially left for later, and were not raised to top priority until the middle to late 1990s.

The most radical market-oriented industrial reform, which was to have the most far-reaching impact, was the establishment of four special economic zones (SEZs) in coastal cities in 1979. Foreign investors in these zones were offered preferential tax and other financial incentives, an open labor market, and duty-free entry for equipment and components for export. The government also invested heavily in requisite infrastructure. Deng restricted the initial scope of the initiative in the face of strong ideological opposition from party leaders.[5] The four SEZs were presented as isolated enclaves that would not contaminate the overall socialist economy. They were also viewed, however, as "schools" to demonstrate how market reforms could work, and the positive demonstration effect was dramatic. In Shenzhen, the largest of the four, from 1980 to 1984, manufacturing firms increased from 26 to 500, the population grew more than 10 times to 350,000, and per capita income rose to almost $2,000, or 7 times the national average. Other cities and provinces quickly sought SEZ status, and by 1984 there were 14 special zones. The number rose to 100 by the end of the decade, and finally, in 1991, the entire Chinese economy was open to foreign investment on the same terms.

These initial reforms stimulated rapid export-driven economic growth during the 1980s. They also, however, created a number of difficult adjustment problems and imbalances within the economy, principally because of the inherent conflicts between the socialist and market-oriented sectors of the overall economy, controlled by an increasingly powerful government bureaucracy. These conflicts deepened over time and have now reached a critical juncture, as described in subsequent chapters. Nevertheless, Deng's initial market-oriented reforms, mostly implemented over a two-year period, 1979-1981, and only five years after Mao's death, were indeed astonishingly rapid and they did fundamentally change the orientation of the Chinese economy from the socialist to the capitalist model.

---

[5] Some economists later criticized the targeted SEZ approach, comparing it with opening the entire Chinese economy to foreign investment, as having impeded Chinese development, but their valid theoretical arguments ignored the political forces in play in 1979. See Edward M. Graham, "Do Export Processing Zones Attract FDI and Its Benefits: The Experience from China," *International Economics and Economic Policy* (Spring 2004), pp. 87-103. Graham chooses not to identify the analysts who believed that the SEZs impeded Chinese development, while noting that "this view is quite prevalent at certain organizations, e.g., the World Bank," p. 94.

The Indian Launch

In May 1991 Rajiv Gandhi was assassinated, and his Congress Party won the ensuing elections largely out of sympathy for its slain leader. The party then chose elderly Narasimha Rao, who was about to retire from Parliament, as the stop-gap leader of a minority government. Not much was expected from the rather dull prime minister who had never held a senior economic position, but Rao surprised everybody by assembling a highly capable economic team, led by finance minister Manmohan Singh, and directing it to design and implement radical changes in economic policy during the summer of 1991.[6] These market-oriented reforms constituted a rejection of India's inward-directed socialist model, developed over four decades under the revered leadership of Jawaharlal Nehru and Indira Gandhi, with deep roots throughout the national political structure.

The trigger for radical change was financial crisis, brought about by relatively poor economic performance, heavy short-term borrowing abroad for several years, surging oil import prices from the first Gulf War, and capital flight by nonresident Indians. Foreign exchange reserves were down to two weeks of imports, and further borrowing from the International Monetary Fund (IMF) was foreclosed without major changes in Indian policy. In addition, growing public disenchantment with the Nehru/Gandhi socialist model had been growing during the 1980s, particularly as the economy of regional rival China grew at 10 percent per year compared with India's 3.5 percent, which was barely above population growth. Rajiv Gandhi had taken some steps toward market-oriented reforms during the 1980s, and the 1991 reforms, in effect, constituted a sharp acceleration of these beginnings.

The resulting reforms turned out to be more rapid and radical than anyone, including the IMF, had anticipated. During the initial weeks, the currency was devalued by 20 percent in the context of transition to a floating exchange rate convertible on current account. Export subsidies and import licenses were abolished, while industrial licenses were reduced in scope, including to permit foreign companies to hold a majority interest in their Indian subsidiaries. Over the next two years, tariff peaks were reduced from 200 to 40 percent, corporate tax rates came down from 56 to 40 percent, and Indian companies were permitted to borrow abroad. Important sectors were freed from the Monopolies and Restrictive Trade Practices Act, a cornerstone of the Indira Gandhi state-controlled system. This ended public sector monopolies in

---

[6] A detailed account of these events is contained in Gurcharan Das, "The Golden Summer of 1991," *India Unbound* (Alfred A. Knopf, 2001), pp. 213-227, from which this summary draws heavily.

banking, airlines, electric power, petroleum, and cellular telephones. As a result, from 1991 to 1993, the central government deficit was reduced from 8.4 percent to 5.7 percent of GDP, inflation declined from 12 percent to 6 percent, and international reserves rose from $1 billion to $20 billion. Foreign direct investment, insignificant at $150 million in 1991, increased steadily to $3 billion by 1997, and the lackluster 3.5 percent growth of the 1980s rose to 5.6 percent during the 1990s.

The reform program was not completed, leaving important gaps, and it lost momentum after the first two years. The rise to center stage of the advanced technology software and other business services sector did not begin until later in the decade. The summer 1991 launch of the economic reform program was nevertheless a radical and far-reaching change in economic course, implemented by a bold and visionary economic team in the face of deeply entrenched opposition from the political establishment and the formidable government bureaucracy known as the "license Raj."

The Parallel Forces in Play

These were the dramatic beginnings of the market-oriented, open trade reforms that not only fundamentally changed the economic structures within China and India, but are also having far-reaching impact on the global economy and the international political order. Five principal forces were in play in both countries, interacting decisively to produce the two successful economic reform outcomes. These same forces, moreover, although in different form, continue to play an important role. They are:

1. *Financial and economic crisis.* The financial and economic crises in China in 1978 and India in 1991, the former more internal and the latter external, forced the leadership to adopt radical change to avoid severe economic setbacks. The sight of the financial gallows concentrated the political minds. In fact, the leadership in both countries used the immediate financial and economic pressures to adopt even bolder market-oriented reforms than would have been necessary.

2. *The failure of the socialist economic model.* In both countries, the inwardly directed socialist model had reached a dead end. The human and economic tragedy of the Great Leap Forward and the Cultural Revolution in China, and four decades of lagging economic performance in India, made the need for fundamental reform painfully evident. Moreover, the ideological champions of communism in China and state socialism in India had no credible alternative model for restoring financial stability and job-creating growth.

3. *The demonstration effect of market-oriented reforms in neighboring economies.* The reform leaders in both countries did not have to look far to see a successful alternative to the failed socialist model. It was the open-trade, market-oriented approach successfully demonstrated within the East Asian region. For China, the salient models were Singapore, South Korea, and, within Greater China itself, Hong Kong and Taiwan. By 1991, India had not only an additional 13 years to observe the Four Tigers, but faced the even more momentous industrial modernization in China, the rival giant power next door. The launching of market-oriented reforms was not the result of one economic model being judged superior to the other in theoretical terms, but of the irrefutable positive demonstration effect in neighboring economies.

4. *Deeply entrenched opposition to reforms within the political establishment.* This was the most important obstacle to more broadly based reform implementation. The economic reformers, although in leadership positions, constituted a minority within communist and socialist political structures with deep historical and ideological roots. The leaderships therefore had to proceed step by step, with caution, which led to uneven implementation and costly gaps. Deng's step-by-step approach for SEZs has already been noted. Rao and Singh launched radical reforms while describing them largely as adjustments to the existing Indian socialist model. They never made a strong public case for market-oriented reforms as an alternative to the failed socialist model and continued to extol the central role of subsidized small farms and businesses as the backbone of the economy. One consequence was that the public did not understand the striking success of the reforms, which contributed to the Congress Party's loss in the election of 1996.

5. *Bold and visionary leadership.* This was critical for the success of the economic reform programs. Both Deng and Singh understood the need for comprehensive, market-oriented reforms, and had a clear vision as to how to proceed to that end. This vision was based on deep-seated conviction and confidence that the desired results would be achieved. This common vision may appear surprising in view of the diverse backgrounds and political orientations of the two men, but their similar personal roots undoubtedly influenced their thinking. Both came from middle-class families that fostered productive work, individual achievement, and deep distrust of heavy-handed, often corrupt, public sector economic enterprises. Both had lived in Europe during their formative years, Deng in France and Singh in England, observing the positive workings of private sector-oriented economies. And both had spent most of their working lives within the power structures of the failed socialist models, thus fully understanding the

need for fundamental change. To borrow Lenin's phrase, they knew what had to be done.

The two leaders also had important differences, of course, which influenced the outcome, and constitute a fitting comparative note on which to end this historical interlude. Singh had a Ph.D. in economics from Oxford and was a distinguished international economist, rising to head the Indian Central Bank. His reform leadership was based on intellectual force and public trust, while relying on political support from Prime Minister Rao. Deng, in contrast, had little formal economic training and rose within the party through forceful political maneuvering and stealth. He was twice purged as a "capitalist roader," and seized control through fierce political struggle after Mao's death.

The most important difference between the two leaders was in their political orientation. Singh's commitment to democracy was reinforced by his belief in a market-driven economy based on private property and the rule of law. Deng, in contrast, was an unyielding advocate of Communist Party authoritarian rule, an advocacy that was in deep inner conflict with his liberal economic agenda. His economic reform principles were reflected in frequently cited quotes: "Seek truth from facts,"[7] an objectivist, scientific-method repudiation of blind faith in Maoist and other communist dicta; and "to get rich is glorious," and "some will become rich faster than others,"[8] which lionized profit making and just rewards from productive achievement. These quotes indicate someone who, in philosophic terms, could be characterized as a "rational individualist." And yet this same leader had been a brutal enforcer of the Great Leap Forward and ended his career by having the troops open fire on Tiananmen Square.

* * *

These same forces continue to interact within China and India, although in some cases in greatly altered form, and constitute the analytic framework for much of the ensuing presentation. The following observations on the current situations in the two countries serve as context for the later analysis:

1. *Financial and economic crises.* The prospect of external financial crisis as a catalyst for further reform no longer exists.

---

[7] This line was adopted by the decisive Communist Party Central Committee meeting in December 1978, orchestrated by Deng. See Online under Deng, Kwan Ha Yim, ed. (Facts on File, Inc., 1991), p. 1.

[8] See Todd Crowell and Thomas Hon Wing Polin, *Asian of the Century, Politics and Government, Deng Xiaoping* (www.asianweek.com, 1999) p. 1.

International reserves of $1.5 trillion by China and $300 billion by India are more than adequate to cushion short- to medium-term financial shortfalls. A domestic economic downturn, however, is possible in both countries, which could trigger further reform actions. For India, a slowdown in growth from 9 to 6 percent, although unlikely, would strengthen the political hand of those pressing for further deregulation, privatization, and other job-creating reforms. A far more serious threat of reduced growth is possible in China if the inevitable adjustment away from export to domestically led growth is poorly managed through counterproductive state controls and a nonresponsive banking system. This outcome is, indeed, likely to happen, and could result either in further reforms to achieve economic recovery or a political impasse with uncertain political consequences.

2. *The failure of the socialist economic model.* This hardly remains an issue since only small extremist groups are calling for a return to full state control of the economy. North Korea and Cuba are not credible role models.

3. *The demonstration effect of market-oriented reforms.* This positive influence on the reform process has strengthened greatly as the demonstration effect has shifted from observed experience in neighboring economies to on-the-ground results within China and India. The most productive and attractive job-creating investments are clearly in the private sector. Continued rapid job creation and growth are also broadly understood to depend on a continued course of market-oriented reforms.

4. *Deeply entrenched opposition to reform within the political establishment.* Such opposition remains in both countries, within the Chinese Communist Party and by communist and other anti-reform parties in the Indian coalition government. The overall balance of political interests in both countries is shifting toward the reform constituencies, through the rapid expansion of the middle class and the growing economic power of the private sector. Opponents of reform in both countries, however, while unable to present a credible socialist alternative, are still able to block further reforms and to achieve rollbacks of some reform actions, such as privatizations. Such actions could lead to confrontation if further reforms become necessary to continue high economic growth and the ideological opponents of reform become fearful of losing their political power.

5. *Bold and visionary leadership.* The leadership factor remains central for further reform, but it has changed greatly, particularly in China. Manmohan Singh returned as Prime Minister of India in 2004, but in a coalition government that includes the communist party and other opponents of reform. He and fellow reformers have retained their visionary goals for reform, but have had to pursue them incrementally,

with caution and largely through stealth rather than boldness. The cumulative reform results, however, have been substantial. In China, the leadership contrast with past performance is far sharper and consequential. President Hu Jintao's leadership style is highly risk averse and accommodating to reform opponents, while his own commitment to further reform is incremental at best. His is the antithesis of Deng's clear and forceful economic reform leadership, and this change casts a dark cloud over the ability of the Chinese government to take difficult economic decisions ahead.

Principal Analytic Findings

The principal analytic findings of the study are:

1. China and India have both been growing at 8 to 11 percent per year during the past four years, and they are now the second and fourth largest economies after the United States, based on the PPP measure. In 2007, the Chinese economy was about 50 percent of the size of the U.S. economy, while India passed Germany to become number four behind Japan.

2. Both countries are pursuing economic strategies of technology-driven industrial modernization open to international trade and investment. China launched its strategy in 1995 and is now a fully engaged advanced technology superstate as defined in this study. India is five to ten years behind China, but is clearly an emerging advanced technology superstate. There are important differences, however, in how the two strategies are being implemented, which will likely affect the course ahead, largely to the relative benefit of India. Three key performance benchmarks are research and development (R&D), foreign direct investment, and exports.

3. China is far ahead of India in R&D expenditures, but both are experiencing rapid growth. From 2000 to 2006, R&D expenditures increased by 19 percent per year in China and 11 percent in India, compared with 4 percent in the United States, the EU, and Japan. India's R&D is still less than a third of China's and is more concentrated in lower-performing public sector programs. Changes over the past couple of years, however, are making India a more attractive location for foreign firm R&D. The 2006 Chinese strategy of indigenous or independent innovation, intended to favor Chinese over foreign firms, together with inadequate protection of intellectual property, cause foreign firms to concentrate R&D on product development for the Chinese market. R&D programs in India, in contrast, are increasingly parts of globally integrated R&D strategies, such as those of General Electric Company (GE), Cisco Systems, Inc., and IBM Corporation.

4. Chinese export-led industrial modernization has been driven principally by foreign direct investment (FDI), which soared in the mid-1990s to reach highs of $70 billion in 2005 and 2006. Indian FDI was only $4 billion to $6 billion annually through 2004, but then surged to $19 billion in 2006 and is projected to be higher yet in 2007, measured on a comparable basis to include reinvested earnings. FDI in China now appears to have leveled off and is perhaps beginning to decline. The FDI surge in India, in contrast, is likely to continue as the result of recent developments, including the 2006 law to liberalize special economic zones, improved investment opportunities for infrastructure projects, and competition among states to ease regulations so as to attract job-creating investment.

5. Export performance is the leading-edge indicator, and both nations are experiencing 25 to 30 percent annual growth, with China again far out front. In 2000, U.S. exports of manufactures were almost three times larger than Chinese exports—$624 billion versus $224 billion—whereas in 2006 China passed the United States to become the number one exporter, with $916 billion versus $786 billion for the United States. Indian exports are far more concentrated in business services, 68 percent as large as manufactured exports in 2006, compared with only 4 percent as large for China, but Indian manufactured exports have begun to take off as well. Chinese exports have shifted definitively to high technology industries, which now account for well over half of total exports, compared with a 20 percent and declining share for labor-intensive industries such as apparel, footwear, and toys.

6. Looking ahead for export performance, India appears well positioned to continue 25 to 30 percent annual growth, while two dark clouds threaten the prospect for China. The first is the very high share of exports by foreign firms, which will be less favored under the indigenous innovation strategy. Fifty-eight percent of total Chinese exports are by foreign firms, which rises to 88 percent for high technology industry exports. The second and far larger cloud is the burgeoning Chinese trade surplus, which rose by $85 billion in 2007, and cannot continue to rise. A major revaluation of the Chinese yuan is inevitable in order to level off and begin to reduce the surplus. This will greatly restrain export growth, beginning with labor-intensive industries. One of the largest beneficiaries of higher priced Chinese exports will be India.

7. The impact of very high growth in Chinese and Indian global trade on U.S. bilateral trade has been a combination of rapid U.S. export growth and even more rapid growth in bilateral trade deficits. From 2004 to 2006, U.S. merchandise exports grew by 60 percent or more to both countries, while the deficits, in 2006, rose to $233 billion with China and $11 billion with India. A particular concern has been

the very rapid increase in the U.S. deficit with China for advanced technology products, the products with the highest R&D and engineering content, from $12 billion in 2002 to $55 billion in 2006. U.S. imports from India of business services, such as software application and engineering, are also a concern for maintaining U.S. technological leadership, but U.S. trade statistics are woefully inadequate, understating imports from India by a likely factor of ten.

8. A longer term objective of both countries, as already noted for China, is to develop indigenous multinational corporations and innovation. Such companies and innovation will develop over time, but with uncertain results during the next several years. Performance varies by sector, with substantial differences between the two countries. Chinese multinational companies are much stronger in the energy and raw materials sectors, while Indian companies are ahead for business services, pharmaceuticals, and probably automotive parts. As for innovation, both countries have advanced programs for aerospace, linked to defense programs. Chinese innovation is noteworthy for nanotechnology, Indian for applied software, and both have expanding results in the biotechnology sector.

9. The growth projection for the coming two to five years is that India is highly likely to continue 8 to 10 percent annual growth, while China is highly likely to experience reduced growth, perhaps to 5 percent to 7 percent, for at least a couple of years, from a "hard landing" transition to more domestically oriented growth.

10. The Indian high-growth projection faces potential obstacles, principally for infrastructure and education, but recent progress in these areas is promising. On the positive side, domestic savings and investment have both risen to above 30 percent of GDP over the past several years to achieve 9 percent balanced growth, with highly entrepreneurial private sector leadership gathering momentum. At the same time, economic reforms are moving forward, albeit slowly, and more by stealth than public decree.

11. The more pessimistic Chinese growth projection stems from the fact that about half of the 11 percent growth in 2007 was from the increase in the trade surplus and related investment in export industry and infrastructure. This growing surplus will have to reverse over the next several years to a progressive decline in order to achieve a reduction in the $500 billion or more of annual central bank purchases of foreign exchange. There are major obstacles to a transition to higher domestic growth, however, including a dysfunctional banking sector, the absence of the rule of law, pervasive corruption, and an extremely low level of personal consumption, which was only 37 percent of GDP in 2007. Transitional adverse trade impact on some manufacturing industries could also lead to political stress within an already risk-averse government unable to take necessary bold decisions.

12. In broader geopolitical terms, China has risen to become the dominant economic power and hegemon in East Asia, while India is in a similar position in South Asia. Moreover, the two regions are becoming more deeply integrated into a single Asian region, with the two subregional advanced technology superstates the principal driving forces. It is unclear whether this bihegemonic relationship within Asia will be of a collaborative or adversarial nature, a question that has important policy implications, including for the United States.

13. In geostrategic terms, Chinese military modernization is progressing steadily to make China the number one military power in Asia, approaching number two globally after the United States. This rise is provoking reactions throughout Asia, including an accelerated modernization of the Indian military, based, as in China, on deepening integration between defense and civilian advanced technology industries. One early military focus of the new Asia-Pacific triangle of the United States, China, and India will be the interaction of their three blue water fleets in the Pacific and through the Malacca Straits to the Bay of Bengal.

14. The longer, 10 to 20 year outlook, is for China and India, as advanced technology superstates, to continue to strengthen their dominant economic and military power positions within their regions. India, however, will also progressively narrow the gap between them, because China, well before India, will reach a more mature stage of industrial modernization, with a consequently slower pace of growth.

15. The longer term prospect in broadest terms is for the world political and economic orders to become increasingly dominated by the four advanced technology regions of North America, West Europe, East Asia, and South Asia, driven principally by the four regional advanced technology hegemons of the United States, the EU, China, and India. In 2006, these four regions accounted for 71 percent of the global population, 82 percent of GDP, 83 percent of exports, 89 percent of military expenditures, and 95 percent of R&D. Looking ahead 10 to 20 years, one can expect all of these percentages, except population, to rise significantly further.

Principal Policy Conclusions and Recommendations

Given these analytic findings, the principal policy conclusions and recommendations are:

1. The advanced technology race between China and India, with both nations on course to become advanced technology superstates, is producing a fundamental transformation in the global economic and

political orders, with far-reaching policy implications for the United States. This technology-driven transformation should bring very large mutual economic benefits to China, India, the United States, and the rest of the world, as a result of highly dynamic gains from international trade and investment. For the benefits to be mutual, however, the international economic system needs to be fair, balanced, and based on market forces rather than government manipulation. The United States also needs to have a domestic economic policy framework that provides incentives to U.S. firms in balance with those provided by its principal trading partners to their firms.

2. Unfortunately, these conditions do not currently prevail. Major distortions within the international economic system, particularly concerning exchange rates, puts American companies at a substantial competitive disadvantage, while the domestic policy framework does likewise in key respects. As a result, the United States has been losing export competitiveness, particularly in the manufacturing sector, which recorded a $630 billion trade deficit in 2006. Moreover, manufacturing and related business services account for over 90 percent of civilian R&D, and thus the trade deficit also poses a risk to long-standing U.S. leadership in technological innovation. The President's Council of Advisors on Science and Technology and the National Academy of Sciences have both warned of the erosion under way in U.S. technological leadership.

3. The United States needs a forceful and comprehensive policy response over the next several years to restore export competitiveness and retain leadership in technological innovation. The starting point for such a response is a sense of national purpose in meeting the challenge, which is sadly lacking. The subject was barely addressed during the presidential candidate debates of 2007 and has not achieved visibility in public opinion polls. This contrasts sharply with China and India, where statements by political leaders and public support for information age economies and technological innovation are ubiquitous. Building a U.S. sense of national purpose should be a top priority for the new administration in 2009.

4. The policy response needs to include various components of international economic policy and the corresponding domestic economic policy agenda. Other areas of national security and foreign policy are also in play, but at this point they are not as deeply engaged with China and India. The policy response offered here thus focuses on the economic relationships, although other issues are broadly considered, including the final point in this summary concerning democratization in China.

5. The international economic policy challenge encompasses international finance, trade, and investment. The most important immediate

issue, by far, is exchange rate policy, and currency manipulation by China and some other Asian nations in particular, whereby exchange rates are maintained far below market-based levels through protracted, large-scale purchases of foreign exchange by central banks. Such currency manipulation violates IMF and World Trade Organization (WTO) obligations, and the United States, together with Europeans and others suffering the adverse trade effects, should pursue a prompt phase-out of currency manipulation both within these organizations and directly with the manipulators. For China, a three-year transition should begin with a minimum 20 percent revaluation of the yuan[9] the first year, subject to two performance criteria: substantial reductions in central bank purchases and the current account surplus which is triggering the purchases. If the criteria are not met, further substantial revaluations would take place in years two and three. If China resists such a transition to more balanced growth, the United States and others should seek recourse through WTO dispute settlement procedures.

6. Trade relationships are more complex, with a mixed picture. Trade is growing rapidly as a result of unilateral reductions in trade barriers and the proliferation of bilateral and regional free trade agreements (FTAs), while the multilateral WTO Doha Round is at a six-year impasse. The United States, bilaterally, should pursue redress or changes in trade practices, particularly by China, that are not in conformance with WTO obligations. The broader medium-term trade objective should be multilateral free trade for the dominant nonagricultural sector, with more modest trade liberalization objectives for trade in agriculture and services. In effect, this objective would involve consolidating the many existing preferential FTAs into a nondiscriminatory multilateral free trade framework. The first stage would be the negotiation of further FTAs, where feasible, as building blocks toward the multilateral free trade objective. The top U.S. priority should be further FTAs across the Pacific, in order to head off an otherwise East Asian free trade grouping that would exclude the United States.

7. International investment policy faces new and daunting challenges from the skyrocketing level of sovereign investment funds, whereby governments enter private investment markets by using their rapidly accumulating central bank reserve holdings. This is thus largely a further consequence of the currency manipulation issue. China, by far, is the largest potential sovereign investor, with reserve

[9] The Chinese currency is called both the yuan and the renminbi. The term "yuan" is used here, as it is in the English version of the annual *Chinese Statistical Yearbook*. "Renminbi" can be difficult to pronounce, easy to misspell, and no other major currency name has three syllables.

holdings headed to $2 trillion in 2008 and $3 trillion in 2010. China is also the country most likely to use such investment power for political rather than purely financial objectives. The initial step in response by recipient nations, principally the United States and the Europeans, would be to require full disclosure of sovereign fund operations. Important operational issues in need of urgent, in-depth examination are screening procedures for mergers and acquisitions, disciplines for equity investments, including for abrupt shifts among national markets and membership on corporate boards, and restraints on government subsidization of international investment, similar, in concept, to the WTO export subsidy agreement and countervailing duties for trade.

8. These various international economic issues together amount to deep disarray within the existing IMF and WTO structures, and a detailed assessment and recommendations for longer term restructuring of these institutions are contained in the annexes to Chapters 8 and 9. An overarching dimension of such restructuring will inevitably be a much larger leadership role for Asia. The U.S.-China-India relationship, in particular, will be decisive. If the three can agree on specific reforms, the reforms will happen, and if they cannot agree, they will not happen.

9. The proposed domestic policy agenda is wide-ranging but presented in less detail. Improved education for science and technology is the top priority, from K-12 and on through university training. Other important policy areas include public spending for R&D in basic research, corporate tax reform, and tort reform. All of these issues need to be addressed in relation to what is happening abroad, including in China and India, where recent policy changes have often adversely affected U.S. international competitiveness. International economic impact statements are proposed for U.S. policy in several areas.

10. Other policy areas, including energy, the environment, and foreign policy, are also addressed but in broader terms. The United States, China, and India are not currently deeply engaged in these areas, except for U.S.-Chinese joint efforts to end the North Korean nuclear weapons program. Specific issues for longer term collaboration—or potential conflict—are, however, identified.

11. The longer term projection, introduced in the final analytic finding, that the global economic and political orders will increasingly be driven by the four advanced technology superpowers—the United States, the EU, India, and China—points to what will likely be the most decisive factor in the future course of world events: democratization in China. The United States, the EU, and India are democracies, based on the rule of law and individual rights, and greatly different scenarios will unfold depending on whether China joins the other three to provide joint international leadership based on democratic values, or remains

the odd authoritarian superpower out. A risk-averse Hu government constantly refers to democratization within China, but has not yet defined the next steps. The coming two to five years, moreover, are likely to be a period of difficult economic restructuring and consequent political stress within China, which is the kind of setting within which political change occurs. The United States should thus place much higher priority on encouraging and supporting democratic change in China, not by threats or economic sanctions, but by more deeply engaged, high level dialogue with the Chinese government and broader public support for democratic forces within China. The first talking point for such frank dialogue should be to state that one-party democracy is the ultimate oxymoron.

Annex 1

The Critical Metric: Purchasing Power Parity Versus Exchange Rate

A critical decision for any comparative assessment of Chinese and Indian economic performance, especially as related to the high income economies of the United States, Europe, and Japan, is what basis to use for measurement, the "purchasing power parity" measure or the "official exchange rate" measure. The choice of measure produces greatly different results for such key indicators as gross domestic product, per capita income, and of particular importance to this study, research and development and defense expenditures. Reporting on relative economic performance, however, often does not identify which measure is used, and there can be inconsistencies in shifting from one measure to the other. The result is considerable confusion and grossly misleading conclusions when the inappropriate measure is utilized. It is therefore essential that the measurement issue be addressed up front, as an annex to this introductory chapter.

The exchange rate measure calculates the value of the goods and services in question at domestic prices and then converts the total into dollars at the official exchange rate. The PPP measure, in contrast, adjusts the values to take account of the generally much lower prices for comparable goods and services produced in China and India, for example, compared with the United States and other high income economies. A simple example is that of a haircut in a first-class hotel in New Delhi compared with a comparable haircut in New York and Tokyo. The haircut might cost five times more, in dollar terms, in the latter two cities, but with the exchange rate measure, five identical haircuts in New Delhi would equate to only one in New York or Tokyo, leading to a fivefold understatement of the output of haircuts in India compared with the United States and Japan. The PPP measure, in contrast, adjusts the value to equate one haircut in New Delhi to one comparable haircut in New York or Tokyo.

The PPP measure is thus, conceptually, the appropriate one for calculating quantities of comparable goods and services produced among nations. This is generally recognized, as stated by Fred Vogel, Manager of the International Comparison Program (ICP): [10] "Clearly

[10] Fred Vogel, "Purchasing Power Parities: Statistics to Describe the World" (ICP, January 2005), pp. 1-2, available at www.worldbank.org/data/icp. The ICP develops comparative PPP and exchange rate measures, as described below for the ICP-2005 survey. It is a joint venture of the United Nations and the University of Pennsylvania, with contributions from the Ford Foundation and the World Bank.

the use of exchange rates gets both the level and changes of the productive capacity of countries wrong. . . . The long-standing recognition of these deficiencies [i.e., of the exchange rate measure] led to the development of Purchasing Power Parities (PPP) as a more appropriate currency converter to compare the GDP and its components across countries."

The use of the exchange rate measure, as in the example of the haircut, thus greatly understates the relative level of output in lower income compared with higher income countries. This understatement is most starkly demonstrated by the comparison of the two measures for levels of GDP, as shown in Table 1-A-1, for 2007. Using the PPP measure, Chinese GDP in 2007 was 50 percent that of the United States, compared with 21 percent using the exchange rate measure. Thus Chinese GDP is approximately 2.5 times larger using the PPP measure. The PPP adjustment is of even greater consequence for India, three times larger relative to the United States. Compared with Chinese GDP, Indian GDP rises from 33 percent using the exchange rate measure to 42 percent using the PPP measure, while for Japan the increase for India jumps from 19 percent to 68 percent.

Table 1-1
Alternative Measures of Gross Domestic Product (GDP)
(2007, $ trillions)

	PPP Measure	**Exchange Rate Measure**
EU (27)	13,812	14,525
United States	13,128	13,128
China	6,571	2,765
Japan	4,106	4,826
India	2,781	926

Source: *2005 International Comparison Program: Preliminary Results*, ICP/World Bank, December 2007. The 2005 figures from this report were projected to 2007 based on actual real growth rates in 2006 and 2007. The figures are thus in terms of 2005 dollars.

The PPP versus exchange rate choice also produces greatly different results for assessing changes over time, in terms of recent trends and future projections. The appropriate choice for measuring recent trends is clearly PPP because the exchange rate measure produces swings in relative levels of the indicator measured as a result of exchange rate fluctuations unrelated to underlying trends. For example, using the exchange rate measure, Japanese GDP as a percentage of U.S. GDP fluctuated from 52 percent in 1990 to 60 percent in 1996 to 38 percent in 2002, largely as a result of changes in the exchange rate. The percentages based on PPP, in contrast, were 38 percent, 39 percent, and 34 percent, respectively, or a more moderate decline reflecting faster U.S. growth during 1996-2002, as well as lower overall percentages from the PPP adjustment for the relatively higher prices for comparable goods and services in Japan.

Both measures have serious biases when projecting current growth rates over the longer term, especially for high-growth, low-income economies like China and India. The biases, moreover, are in opposite directions, but for the same reasons. PPP projections overstate future levels because the PPP differential tends to decline over time as a result of currency appreciation and/or relatively high inflation, which characterize such high-growth economies. The exchange rate measure understates future levels because the projections likewise do not take account of such currency revaluation and/or inflation. For this study, the PPP measure is generally used for both recent trends and future projections, as related to the indicators enumerated below, but projections are limited to the relatively short term of two to five years, wherein the PPP overstatement should be relatively small.

The principal problem with the PPP measure has been that it is an estimate, and PPP estimates have been of poor quality, not fully compatible, and out of date, including for China and India. Fortunately, this long-standing problem has now been largely resolved through the results of the ICP-2005 survey of December 2007.[11] For the first time, a comparable set of over 1,000 individual goods and services has been compared for 146 countries, based on data collected in 2005. The project was coordinated by the World Bank and integrates the PPP work of the five ICP regional programs for developing countries with the Eurostat/OECD program for the developed countries. ICP-2005 regional results were reported during 2006 and 2007, and the preliminary composite results were released in December 2007. These results were used in Table 1-A-1.

The ICP-2005 results are important for two reasons. The first is that the PPP estimates show a much smaller adjustment compared with the exchange rate measure than did the earlier PPP estimates. For China, the earlier, poor quality PPP adjustment increased Chinese GDP four times compared with U.S. GDP, while the ICP-2005 adjustment increases Chinese GDP two and a half times. The second and more lasting reason is that, at last, there is a reasonably accurate set of PPP estimates. As these estimates become more widely known and

---

[11] 2005 *International Comparison Program: Preliminary Results*, ICP/World Bank, December 17, 2007.

understood, they should reduce the misleading statements from the inappropriate use of the exchange rate measure by government officials, the media, and other observers.

The approach adopted in this study for the selection of the appropriate measure is based on four guidelines:

1. A choice has to be made between the two measures. The issue is too important to be hedged on an either/or basis.
2. The appropriate choice depends on what question is being addressed. In some cases, the appropriate choice will be the PPP measure, and in others it will be the exchange rate measure.
3. When the question concerns comparative economic performance based on quantities of comparable goods and services produced, the appropriate measure is PPP. Such questions predominate throughout this study, as explained below.
4. Although the PPP measure is an estimate based on best available information, this is not grounds for rejecting it, as some critics do. The fact is that the exchange rate measure, for a number of key issues, greatly understates the relative performance of China and India compared with the United States and other high income countries, and is thus highly misleading. The appropriate response to such critics is the sage advice attributed to John Maynard Keynes: It is better to be approximately right than precisely wrong. In any event, the far better ICP-2005 PPP estimates greatly diminish the rejectionist argument and make the PPP measure a much closer approximation of the truth.

In this study, four principal indicators are consistently measured on a PPP basis: GDP, per capita income, R&D expenditures, and defense expenditures. Two others—current account balance as a percent of GDP and size of export markets—are measured on the exchange rate basis, but with qualification. Comment on each of the six indicators is as follows:

Gross domestic product (GDP).—Both measures are commonly used for this most frequently cited comparative indicator. PPP is clearly the appropriate measure, but the large majority of references, particularly in the media, use the exchange rate measure. If the basis of measurement is not cited, one can assume it is the exchange rate measure, while PPP figures are almost always referenced as such. The exchange rate measure is widely used, in part, because official national statistics, and dollar-based comparisons by the IMF and other organizations, are generally on this direct rather than estimated basis. There are also political reasons why governments choose the exchange rate measure for GDP, which greatly understates the relative size of the Chinese economy, in particular, compared with the United States and other high income countries. China wishes to downplay its relative size to justify its position as a poor developing country, in the WTO and elsewhere, which can qualify China for preferential treatment. Japan prefers the exchange rate measure so as to maintain that it is still the second largest economy. The almost total official use of the exchange rate measure by the United States appears similarly, and even more importantly, to avoid having to confront the fact that the Chinese economy is now half the size of the U.S. economy and on track to be of comparable size by about 2020.

Per capita income (PCI).—Per capita income is a less controversial indicator and is more frequently measured on a PPP basis. The PPP measure is obvious for measuring poverty levels and middle class status, which are usually defined in terms of PCI, and where much lower relative prices in low income countries need to be taken into account. Most of these calculations, moreover, are done by professional economic analysts who understand the measurement issue.

Research and development (R&D).—This has also generally come to be measured on a PPP basis in recent years, in large part through the extensive international comparison of R&D expenditures by the Organization for Economic Cooperation and Development (OECD). An OECD survey of Chinese R&D expenditures relative to the high income economies adopted the PPP measure, with the methodological note that, "The exchange rate is hardly a plausible measure of relative prices and relative price structures between the United States and China."[12] The need for a PPP adjustment for R&D expenditures is also self-evident from the plethora of reporting on R&D outsourcing to China and India, with much lower costs for comparable R&D work a major reason for the outsourcing. For example, the typical cost for development and testing of a new drug in China is estimated at $120 million, compared with $1 billion in the United States, or a ratio of 8 to 1, triple the PPP 2.5 to 1 ratio for GDP.[13] The PPP estimates for R&D, however, have generally been based on the broader GDP estimates, and the more detailed ICP-2005 results should enable the OECD to make more targeted PPP estimates for R&D.

Defense expenditures.—This is an important, essentially U.S.-China bilateral measurement issue. The relative costs for defense industry and military personnel and operations are clearly far lower in China than in the United States, which means PPP is the appropriate measure. And yet both the Chinese and U.S. governments exclusively use the exchange rate measure for comparing defense expenditures, greatly understating the relative size of Chinese military spending.

---

[12] See *An Emerging Knowledge-Based Economy in China? Indicators From OECD Databases* (OECD, 2004). The quote is from page 6.

[13] See *BusinessWeek*, May 28, 2007, p. 61. The source of the estimated cost was Frost & Sullivan, Vital Therapies Inc.

This is understandable for China, which has tried to downplay the size of its military buildup in recent years. The U.S. reason for using the exchange rate measure, including in the Department of Defense Annual Report to the Congress on Chinese military power, is far less clear. It would certainly be more alarming to assess Chinese expenditures at about 50 to 70 percent of the U.S. level, using a factor of 2.5 for the PPP differential, rather than at the 20 to 30 percent level contained in recent annual reports. This defense expenditure measurement issue is discussed further in Chapter 6, with the conclusion that the official U.S. exclusive use of the exchange rate measure is puzzling, to say the least.

Current account balance as a percent of GDP.—This measure, which figures prominently in IMF and other discussion of trade adjustment, is almost always on an exchange rate basis. The basic rationale is that a reduction in a current account surplus or deficit would take place in dollar terms, as would the adjustment effects transmitted throughout the economy. The trade adjustment process figures prominently in this study, in view of the very large and growing Chinese current account surplus in particular. But although the exchange rate measure is justified in the immediate context, it can be misleading in projecting the adjustment process ahead because it does not take account of anticipated exchange rate adjustments related to the current account adjustment, which directly change the ratio of current account balance to GDP. For example, when the Argentine economy was on the verge of financial collapse in 2001, the current account deficit/GDP ratio was about 6 percent, which did not appear overly alarming, but when the Argentine peso then plunged by 50 percent, the ratio rose to 12 percent, which was alarming. This study uses the standard practice of exchange rate measurement for the current account balance/GDP ratio, while recognizing that this would change, in particular, to the extent the Chinese yuan and the Indian rupee were revalued.

Size of export markets.—This is also normally calculated with the exchange rate method, as a measure of the dollar amount of available spending in the export market. Again, however, if exporters are planning ahead several years in developing export markets, they need to take account of likely changes in exchange rates. For example, if the yuan should appreciate by 50 percent over the coming several years, the dollar value of the Chinese market for exports would also rise by 50 percent. This study again follows the standard exchange rate measure for export markets, but with a cautionary note about the effect of anticipated changes in exchange rates.

PART I

The Analytic Assessment

CHAPTER 2

The Rise of Science and Technology to Center Stage

When Chinese Premier Wen Jiabao visited India in April 2005, he predicted: "It is true India has the advantage in software and China in hardware. If India and China cooperate in the information technology industry, we will be able to lead the world . . . and it will signify the coming of the Asian century of the information technology industry."[14] China, moreover, has now placed top priority on technological innovation, as elaborated in another statement by Premier Wen in Beijing, also in April 2005, with "independent" referring to Chinese innovation: "Science and technology are the decisive factors in the competition of comprehensive national strength . . . we must introduce and learn from the world's achievements in advanced science and technology, but what is more important is to base ourselves on independent innovation because it is impossible to buy core technology . . . independent innovation is the national strategy."[15]

Wen sets forth a sweeping vision in an age of rapid technological innovation and application, centered on the information technology sector, and in recent years China and India have made considerable progress toward the advanced technology objectives described by him. The questions remain, however, as to how fast and how far they have progressed to date, and how much further they are likely to go in the years ahead. This chapter begins the assessment by describing the policy frameworks related to advanced technology industry that have evolved since the initial reforms were launched, which is followed by a comparison of the performance in the two countries with respect to three principal resources: research and development (R&D), technology-oriented education, and foreign direct investment (FDI). Chapter 3 then examines trade performance, with a focus on export competitiveness in advanced technology industries, while Chapter 4 addresses the more elusive yet critical objectives of independent, also referred to as indigenous, multinational corporations and technological innovation. Chapter 5 brings all of these developments together in a net assessment of the likely courses ahead for China and India.

---

[14] *Washington Times*, April 11, 2005.
[15] Cited in Michael Pillsbury, "China's Progress in Technological Competitiveness: The Need for a New Assessment" (paper presented to the U.S.-China Economic and Security Review Commission, Palo Alto, CA, April 21, 2005), p. 5.

The Advanced Technology Policy Framework in China

The Chinese economic reform program launched in 1979 set agriculture and industry as the top priorities among the Four Modernizations, with science and technology a distant third, followed by military modernization. The results during the ensuing 15 years bore out this prioritization. Large gains were achieved in farm output and productivity, and light industry flourished, especially in the broadening network of special economic zones. This concentration in labor-intensive industry was reflected in the relatively low level of foreign direct investment despite the attractive tax and other incentives offered. The fact is that labor-intensive industries, such as apparel, footwear, furniture, and toys, do not require much capital investment. The results are shown in Figure 2-1, where foreign direct investment remained below $5 billion per year through 1991. At that point, however, with the entire Chinese economy finally open to foreign investment, higher levels of investment began to flow in, including in more upscale, capital-intensive industries, while at the same time the Chinese government began to focus on science and technology as a top priority within the overall economic strategy.

As early as 1985, the Communist Party Central Committee had made the decision to reform the science and technology management system and give it a strong commercial orientation: "Modern science and technology constitute the most dynamic and decisive factors in the

Figure 2-1
FDI Flows Into China
1980-2006

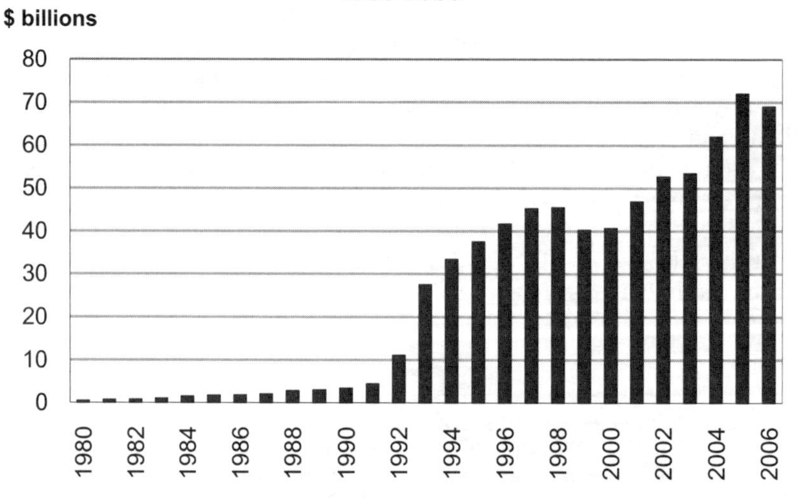

Source: Chinese Foreign Investment Department, Ministry of Commerce.

new productive forces. . . . Our scientific and technological work must be oriented to economic constructions."[16] In 1988, the Torch Program established 53 High Technology Development Zones (HTDZ), emphasizing applied research for commercial application and inviting international partners to collaborate and invest in high technology industries. Implementation of these initial plans were weak, however, and it was only in 1995 that a more specific and action-oriented "Decision on Accelerating Scientific and Technological Progress" was adopted, in part as an initiative by China's new generation of leaders to assert their influence in advanced technology development.[17] Thus, 1995 can be identified as the landmark beginning of the proactive Chinese advanced technology development strategy.

A number of specific actions were taken as a result of the 1995 decision, including rapid increases in advanced education and government spending on research and development. Greater incentives, including export credits and tax rebates, were offered to joint ventures with foreign companies to attract new technology and management skills. The broad objective was to develop advanced technology to "match those of the advanced countries in some fields,"[18] with the information technology sector at the top of the list. During the late 1990s, a series of "Golden Projects" was implemented to create an advanced communications infrastructure, including fiber optic networks in banking services, health and medicine, and tax collection, to form China's "information superhighway."

The most important strategic decision for the development of advanced technology industry during the initial ten years through 2005 was the central role given to foreign direct investment (FDI) for the rapid creation of such export-oriented industry. Foreign investors received more favorable tax and other incentives than Chinese companies, which drew criticism from some observers[19] and induced

---

[16] This rise of science and technology to political and operational prominence in China is described in Kathleen Walsh, *Foreign High-Tech R&D in China* (Henry L. Stimson Center, 2003), Chapters 2-3. The presentation here draws largely from this work.

[17] China suffered a "lost generation" of university-educated leaders during the Cultural Revolution, and the older generation of leaders continued to fill the gap until the mid-1990s, when younger scientists and engineers of the post-1976 generation, including many trained abroad, began to come of age.

[18] From the 1995 decision.

[19] This relationship is developed by Yasheng Huang, *Selling China: Foreign Direct Investment During the Reform Era* (Cambridge University Press, 2003). Huang concludes that if privatization of state-owned enterprises had been politically feasible or if Chinese private entrepreneurs had been able to obtain financing to fulfill export contracts, many of the benefits currently associated with FDI could have been realized with a lower level of FDI.

considerable "round tripping" during the late 1990s, whereby Chinese firms went abroad to Singapore and elsewhere and then returned to China to reap the preferred tax status as foreign firms. As a result, FDI shot up from $4 billion in 1991 to $45 billion in 1997. When FDI declined in 1999-2000, and investors appeared unwilling to bring in production and R&D for more advanced technology products through joint ventures with Chinese firms, approval of 100 percent foreign ownership was greatly liberalized, and the 100 percent ownership share of total FDI rose from 39 percent in 1999 to 68 percent in 2004.

The Chinese experience of foreign direct investment as the critical catalyst for advanced technology development stands in sharp contrast with the earlier experience of industrial modernization in Japan, South Korea, and Taiwan, where FDI was greatly restricted and continues to be relatively small. This greater role for FDI was one reason why industrial modernization moved forward more rapidly in China since 1995 than it did earlier in the other East Asian economies. It also reflected the reality that there was little indigenous capability in China in 1995, with Chinese private companies only recently created and the state-owned enterprises greatly overstaffed and noncompetitive. The most extraordinary result, as explained in Chapter 3, is that 88 percent of Chinese exports in high technology industries are by foreign invested firms.

One other important component of the Chinese advanced technology policy framework is the highly decentralized regulatory and tax structure, with considerable decision-making powers given to provincial governments, including approval of and tax incentives for foreign investors. In 2004, the minimum level of investment requiring central government approval was raised from $30 million to $100 million, thus excluding the large majority of investments in the manufacturing sector. The net result has been fierce competition among the provincial and city governments to attract foreign investors and related high-paying jobs. Such competition has been especially intense in the semiconductor sector, where about 15 foundry plants have been built or are under construction, mostly in Shanghai and Beijing.[20]

The entire process of advanced technology industry development has gone forward in parallel with a broad economic and trade liberalization strategy focused largely on commitments related to the 2001 Chinese accession to membership in the World Trade Organization (WTO), implemented progressively through 2007. The specifics of

[20] For example, Shanghai offered a "five plus five" incentive of a five-year tax holiday plus an additional five years at half the tax rate, and Beijing countered with a "Shanghai plus one" plan of one additional year of tax benefits beyond whatever Shanghai offered.

these trade and trade-related commitments are described in Chapter 9. The overall impact on advanced technology trade and investment has been highly positive, although not without problems, including protection of intellectual property rights, technology-related commitments for investment approval, and technical standards.

This initial period of advanced technology development and export-oriented growth, driven principally by foreign direct investment, is now, however, in transition to a second stage aimed at shifting the concentration of technological innovation and exports from foreign to Chinese companies, the national strategy of independent innovation expressed by Premier Wen in 2005. In December 2006, during the first meeting in Beijing of the U.S.-China Strategic Economic Dialogue, headed on the U.S. side by Secretary of the Treasury Henry Paulson, Chinese Vice Premier Wu Yi restated the strategy: "China will take enhancement of innovation capability as the strategic starting point for scientific and technological development . . . [and] endeavor to achieve breakthroughs in strengthening the ability of independent innovation."[21]

The content of the new strategy is not yet clear and it is controversial within China, but the direction of change is toward greater economic nationalism, with the potential for conflict with WTO and other commitments to provide open markets and national treatment for foreign firms. It is also a high-risk strategy, considering the preponderant role of foreign firms in Chinese exports. Strong preferences for Chinese firms would be a disincentive for continued high levels of foreign investment in China, while alternative locations for export-oriented investment, including India, are increasingly attractive.

Implementation of this new strategy of independent innovation was officially launched in January 2006 through the 15-year "Medium- to Long-Term Plan for the Development of Science and Technology."[22] The plan was three years in the making, with the participation of more than 2,000 scientists, engineers, and corporate executives. Ambitious goals are set for increased R&D in targeted sectors—in energy, manufacturing, aerospace, and biotechnology. There are 17 designated engineering and scientific megaprojects, including advanced numeric-controlled machinery, extra large scale integrated circuits, manned aerospace and moon exploration, new-generation broadband wireless telecommunications, and nanotechnology.

The formulation of this program was controversial with respect to the future role of multinational companies. The debate was described as follows: "Some Chinese economists argued strongly that at China's

---

[21] Vice Premier Wu: China's Development Road, December 14, 2006, GOV.cn.

[22] The background and contents of the plan are described in Cong Cao, Richard P. Suttmeier, and Denis Fred Simon, "China's 15 Year Science and Technology Plan," *Physics Today*, December 2006, pp. 38-43.

current level of economic development . . . the most cost-effective way to upgrade China's technological capabilities would be to continue to encourage technology transfers from multinational corporations. . . . Most members of the technical community rejected that thinking and argued that foreign corporations could no longer be counted on to transfer technologies. . . . They claim that China's technical gains from multinational corporations were disappointing. . . . In addition, China had become increasingly dissatisfied with the relative gains it was accruing. . . . The royalties Chinese firms had to pay for foreign technology . . . often seemed excessive. . . . It is clear that the advocates of a strategic S&T policy to strengthen indigenous R&D clearly have won out."[23]

The specific policy implementation of the new program is moving forward on many fronts, but in an ill-defined manner, reflecting continued controversy. The innovation adjective can be translated into English as independent, indigenous, and home grown, which have different connotations. "Home grown" innovation could include that of foreign firms within China, or at least by joint ventures. "Independent," in contrast, clearly means Chinese and not foreign firm innovation. The effects of the new strategy will become clear as new policies and regulations are adopted and implemented. Several of them are of growing concern to American investors in and exporters to China, as discussed in Chapter 9 related to bilateral trade discussions, including, for example, the 2006 Revitalization of the Industrial Machinery Manufacturing Industries program. Another concern arises from the November 2007 revised guidelines for foreign direct investment which include restrictions on such investment in "strategic and sensitive industries relating to national economic security."

In broadest terms, the economic nationalist direction of the new strategy is clear. Foreign firms are no longer praised for their advanced technology investment and job creation as they were ten or even five years earlier, and official comments now are more apt to complain about excessive profits and the unwillingness of foreign companies to bring more advanced R&D work into China. There is nevertheless great uncertainty as to how fast and how far the government's preference for Chinese over foreign firms will be pushed. China's dependence on foreign investment for advanced technology exports is understood, and WTO obligations are a limiting factor. The darkest cloud overhanging the independent innovation strategy is the difficult adjustment ahead away from export-oriented to domestically oriented growth, which will inevitably involve a substantial revaluation of the yuan and thus have substantial adverse impact on the manufacturing sector, as discussed in Chapter 5. And this adjustment will cut both ways: Stronger demands

---

[23] *Ibid.*, p. 41.

to provide greater support for struggling Chinese firms and growing concern that foreign firms, the backbone of Chinese advanced technology exports, will shift investment elsewhere.

The Advanced Technology Policy Framework in India

An Indian policy framework targeted on the development of advanced technology industry has only recently emerged and is being implemented in piecemeal fashion. The initial reform program of 1991 unleashed the Indian private sector from 40 years of suffocating regulation, taxes, and state monopolies under the socialist rule of Nehru and Indira Gandhi, but major roadblocks remained for industrial modernization, and for technology-intensive industry in particular. Infrastructure was lacking, education was inadequate, labor laws discouraged hiring, many sectors were reserved for small business, and pervasive regulation continued. Unlike China, however, India had a broad structure of competitive private firms, some large and dating back to the 19th century, and they unleashed a quick positive response to the initial reforms. The investment climate for moving up the ladder of applied new technologies through the late 1990s nevertheless remained largely negative, and initial successes in this area were described as "in spite of" rather than "because of" government policy.

The big breakthrough in the late 1990s was in the software and related information technology services sector, where a burst of new investment and job creation was enhanced by lucrative contracts for Y2K computer reprogramming. Export-oriented software firms went multinational at a rapid pace, but more as a result of loopholes in and exceptions by stealth to the government regulatory framework than as a response to conscious government support for such entrepreneurial initiative. The services sector was also unencumbered by physical infrastructure constraints, such as inadequate roads and ports. A decisive action was the provision of independent satellite links for software firms, which was an exception to the highly regulated telecommunications sector and attributed to the personal initiative of a visionary civil servant.[24] Another important exception to the rule was cable TV, which "the license Raj somehow forgot to regulate," leading to billions of dollars of private investment, hundreds of miles of fiber optic cable, and Internet as well as TV access for tens of millions of people.[25]

Still another example of forward-thinking private sector initiative, in spite of government policy, was the creation of the National Institute

---

[24] See Das, *op. cit.*, p. 247.

[25] See *The Economist*, May 27, 2000.

of Information Technology (NIIT). Public sector education was grossly inadequate in turning out university and other graduates qualified in computer and information technology skills, and to rectify this gap two young entrepreneurs, in 1981, formed the private sector NIIT, which grew spectacularly, including through the franchising of schools throughout the country and abroad. The profitable company now has four thousand "learning centers" that have turned out four million graduates with computer-related skills. The public education establishment, however, threatened by the exodus of students from public to superior private schools, resists such private sector initiative, and NIIT has still not received government accreditation.[26]

The Ninth Five-Year Plan, for 1997-2002, included a lengthy statement on "Science and Technology."[27] The statement lauded the software sector, but decried the poor performance of the hardware industry and the fact that Indian industry owned very few patents. It called for the rapid expansion of R&D with a private to public sector ratio of 4:1, an enormous shift from the existing 3:1 ratio in favor of the public sector, but there were no specific implementation proposals for this or other objectives, and the only details presented were for public sector science and technology projects. In 1999, the government formed a Ministry of Information Technology, but observers, including the private sector, were not impressed. The new ministry created a $20 million venture capital fund, but it was the bureaucrats who chose the recipients, and the media referred to it as the "Nephews and Nieces Fund."

Finally, in January 2003, Prime Minister Atal Bihari Vajpayee unveiled a more action-oriented and comprehensive strategy, "Science and Technology Policy 2003,"[28] which his successor, Prime Minister Singh, strongly endorsed in 2004, with the accompanying comment:

> We take satisfaction from the fact that over 100 global companies have come to India to set up R&D Centres, affirming the intellectual capital of our scientific and engineering community. Science must grapple with the key challenges facing the country today. These include the pressures of increasing population, greater health risks, changing demographics, degraded natural resources, and dwindling farmlands. We need new science and technol-

---

[26] See Gurcharan Das, "The India Model," *Foreign Affairs*, Summer 2006, p. 12.
[27] http://www.planningcommission.nic.in/plans/planrel/fiveyr/10th/default.htm.
[28] Department of Science and Technology, Ministry of Science and Technology, http://www.dst.gov.in/stsysindia/stp2003.htm.

ogies, new priorities and new paradigms to address these fundamental challenges.

Among the principal policy objectives and implementation measures contained in this policy statement are the following quotes:

- To encourage research and innovation . . . by promoting close and productive interaction between private and public institutions in science and technology. Sectors such as agriculture (particularly soil and water management, human and animal nutrition, fisheries), water, health, education, industry, energy including renewable energy, communication and transportation would be accorded highest priority. Key leverage technologies such as information technology, biotechnology, and materials science and technology would be given special importance.
- To establish an Intellectual Property Rights (IPR) regime which maximizes the incentives for the generation and protection of intellectual property by all types of inventors. The regime would also provide a strong, supportive and comprehensive policy environment for speedy and effective domestic commercialization of such inventions so as to be maximal in the public interest.
- A certain percentage of the overall allocation of each of the socio-economic ministries [is] to be devoted for relevant programs and activities in science and technology.
- A major initiative to modernize the infrastructure for science and engineering in academic institutions will be undertaken. It will be ensured that all middle and high schools, vocational and other colleges will have appropriately sized science laboratories. Science, engineering and medical departments in academic institutions and universities and colleges will be selected for special support to raise the standard of teaching and research.
- New mechanisms would be instituted to facilitate the return of scientists and technologists of Indian origin to India, as also their networking, to contribute to Indian science and technology.
- Intensive efforts will be launched to develop innovative technologies of a breakthrough nature; and to increase our share of high-tech products. Aggressive international bench-marking will be carried out. Simultaneously, efforts will be made to strengthen traditional industry so as to meet the new requirements of competition through the use of appropriate science and technology.
- Every effort will be made to achieve synergy between industry and scientific research. Autonomous Technology Transfer Organizations will be created as associate organizations of universities and national laboratories to facilitate transfer of the know-how generated to

industry. . . . Industry will be encouraged to financially adopt or support educational and research institutions, fund courses of interest to them, create professional chairs, etc., to help direct S&T endeavors towards tangible industrial goals.

- Efforts by industry to carry out R&D, either in-house or through outsourcing, will be supported by fiscal and other measures. To increase their investments in R&D, innovative mechanisms will be evolved.

This is an ambitious program, and the final two excerpts, in particular, highlight the central role that the private sector needs to play if the program is to succeed. It is noteworthy that prime ministers from both principal political parties endorsed the detailed policy framework, and government leaders are constantly reiterating the top priority given to science and technology within a knowledge-based economy. Implementation of the program has been slow to materialize, however, which is in keeping with the complex workings of Indian democracy under a coalition government and the limitations on financial resources.

The basic strategy for developing advanced technology industries in India, in any event, is not to target large amounts of financial resources for specific mega projects, as in China, but rather to progressively open the Indian market to private sector investment in all sectors, and then let the invisible hand of market forces do the rest.

An important step in this direction was taken in February 2006, when the existing law for the establishment of special economic zones (SEZs) was greatly liberalized. A five-year holiday for the profit tax, exemption from import and excise duties, a substantial reduction in regulatory requirements, and private sector financing of infrastructure within the SEZs, offer great incentives for export-oriented industry. This initiative played out in typical rough and tumble Indian fashion, with the economic reform team in the lead. The initial private sector reaction surprised almost everybody, including the anti-reform politicians and the license Raj. In the first six months, there were 400 applications for and over 200 approvals of SEZs, including two very large, town-sized SEZs near New Delhi and Mumbai by Reliance Industries, one of India's largest companies.

Then opposition built up, especially over compensation for displaced farmers, reaching a peak in January 2007 in West Bengal, where six people died in street protests, with massive media coverage. The communist government of West Bengal strongly supported the job-creating SEZs and responded to the protests with more generous compensation, whereby displaced farmers would receive a fair market value for their land plus a hardship bonus payment plus job training at the SEZ for at least one member of each displaced family. Opposition to the SEZs for various reasons has continued, which led to some restrictive amendments to the law in April 2007, but the overall proliferation of SEZs to spur industrial growth will almost certainly continue. This vibrant and unruly democratic process in India, moreover, is far superior to the Chinese experience, explained in Chapter 5, where farmers are often displaced with little or no compensation, in the context of violence by police and corruption, and without recourse to a free press or independent courts.

Another important initiative in 2006 was the launch of The National Strategy for Manufacturing, strongly endorsed by Prime Minister Singh: "Indian industry must aspire to be a global player in areas like agro-processing, textiles and garments, automobiles and auto components, pharmaceuticals, chemicals and petrochemicals, and leather and footwear."[29] It called for 12 to 14 percent annual growth over the coming decade, up from 7 percent growth during 1995-2004, and by early 2007 growth in manufactures had already reached the 14 percent target. The 80 page strategy report contains a wide range of recommendations in the areas of trade liberalization, tax reform, education, infrastructure, and labor and regulatory reform, which are being implemented, albeit slowly and in piecemeal fashion, with strong support from the private sector.

* * *

Within these two policy frameworks, the rapid rise of advanced technology industry in China and India has centered largely on the interaction of three economic components: research and development, technology-oriented education, and foreign direct investment. Expanding R&D programs drive the development and application of new products and processes, which, in turn, require the employment of large numbers of engineers, scientists, and other technology-oriented personnel. This interaction, moreover, is taking place to a large extent, in both countries, through investment by foreign companies and joint ventures. The pace and structure of the overall interactive process, however, has differed greatly between China and India, and has been evolving at a rapid pace. Each of the three dimensions is examined here, in this dynamic setting, to provide the best current reading of events and to highlight key elements of change that will influence the course ahead.

---

[29] Government of India, *The National Strategy for Manufacturing*, National Manufacturing Competitiveness Council, March 2006.

Research and Development

This is an area where, until very recently, India has lagged far behind China, particularly in the private sector. It is also the most difficult of the three components to assess in qualitative terms. There is no question, however, that since 1995 R&D expenditures have been growing far more rapidly in China and India than in the United States, the EU, and Japan, which until the mid-1990s had dominated global R&D expenditures. Such expenditures over the period 1995-2006 are presented in Table 2-1, measured in terms of purchasing power parity (PPP). The absolute figures in the table overstate R&D expenditures in China and India because they are based on the earlier estimates for the PPP adjustment rather than the ICP-2005 estimates presented for GDP in the annex to Chapter 1. The annual growth figures for 2000-2006 nevertheless show the sharp differential in recent growth: 19 percent for China and 11 percent for India, compared with 4 percent each for the United States, the EU, and Japan.

Table 2-1
R&D Expenditures
($ billions, PPP)

	1995	**2000**	**2006**	**Annual Growth 2000-2006**
United States	184.1	264.0	328.9	4%
European Union (15)	130.8	175.7	224.0	4%
China	18.4	48.3	136.3	19%
Japan	78.7	98.6	127.8	4%
India	10.2*	20.7	38.9	11%

*1996.
Sources: OECD Factbook 2006: Economic, Environmental and Social Statistics for 2000 and 2006; and *An Emerging Knowledge-Based Economy in China? Indicators from OECD Databases* (March 22, 2004), for 1995 data.

Another broad measure of R&D performance is the percentage of GDP spent on R&D, presented in Table 2-2. Again, China shows a strong upward trend, compared with the considerably higher yet relatively flat percentages for the United States, the EU, and Japan. China rose from 1.0 percent in 2000 to 1.4 percent in 2006. The Chinese level was probably up to 1.5 percent in 2007, and the official target for 2010 is 2.0 percent, or higher than the 1.8 percent EU level in 2006. The Indian percentage also rose from 0.8 percent in 2000 to 1.0 percent in 2006, but the upward trend is less clearly established because the percentage had stagnated around 0.8 percent for two decades before dipping to 0.7 percent in 2000 and then rising to a new high of 1.0 percent in 2005-2006. These increases in R&D share of GDP for China and India, together with much higher GDP growth, explain the very rapid rise of relative levels of R&D expenditures for both countries shown in Table 2-1.

Table 2-2
R&D as a Percent of GDP

	2000	**2006**
United States	2.7	2.5
European Union (15)	1.9	1.8
China	1.0	1.4
Japan	3.0	3.1
India	0.8	1.0

Sources: OECD Factbook 2006: Economic, Environmental and Social Statistics (for 2000 data); and R&D Magazine Global R&D Report (for 2006 data).

Table 2-3 breaks down R&D expenditures in 2006 by performing sector: industry, government, and university. There are broadly similar structures for the United States, the EU, China, and Japan. Industry R&D dominates in all four, ranging from 77 percent in Japan to a low of 62 percent in China. China had a consequently higher share of government expenditures, at 27 percent and a somewhat lower, 11 percent share, for universities. The most striking contrast in Table 2-3, however, is the Indian structure of R&D dominated by government expenditures, with 75 percent of the total, and far lower shares for industry, at 23 percent, and universities, at only 2 percent. These differences in the sectoral percentage shares are converted into expenditures in the bottom half of the table, with equally striking contrasts for India, in particular. Industry R&D in India is only 10 percent of that in China and 4 percent of the U.S. level, while university R&D in India is close to negligible.

Within these broad quantitative dimensions of R&D expenditures, a number of qualitative factors influence the results, in terms of technological innovation and increased productivity, but factual information and hard analysis are lacking in many areas. Basic research, for example, with a time lag, can have substantial payoff for the development of applied new technologies, and in this area China and India remain far behind the United States, the EU, and Japan. China, however, recently placed high priority on increased spending for basic research at universities, including through joint ventures with foreign firms, and an evaluation of current performance and outlook in this area would be useful. In the case of India, the limited university resource base indicates a very small amount of basic research, but this could change to the extent the Science and Technology 2003 program is implemented.

Table 2-3
R&D by Performing Sector (2006)

	Percentage of Total R&D		
	Industry	**Government**	**University**
United States	74	12	14
European Union	68	13	19
China	62	27	11
Japan	77	9	14
India	23	75	2

	Expenditures ($ billions)		
	Industry	**Government**	**University**
United States	243.4	39.5	46.0
European Union	163.3	31.2	45.6
China	84.5	36.8	15.0
Japan	98.4	11.5	17.9
India	8.8	28.8	0.8

Source: OECD Factbook 2006: Economic, Environmental and Social Statistics.

Public sector R&D is another area where qualitative assessments are limited. Much of this R&D, especially in the United States and China, is for military modernization, which can have dual-use spinoff benefits for the commercial sector, but much of this research is kept secret and, for China, is off budget. Closely related aerospace public sector R&D is important in all five economies and would be worthy of an in-depth comparative analysis. The agriculture and energy sectors also receive considerable public sector R&D support in both China and India. The bulk of public sector R&D in India takes place in independent research centers, with limited linkages to universities and the private sector. These centers have pockets of excellence, but the general view is that there is a low level of results in terms of new product and process development. Some commentary on these various public sector R&D programs is provided in Chapter 4, but more in-depth analysis, on a sectoral basis, would be highly useful.

The central thrust of R&D programs for developing Chinese and Indian advanced technology industries, in any event, comes from the private sector and this relationship is evolving rapidly, in qualitative as well as quantitative terms, particularly as multinational corporations (MNCs) with the largest R&D programs globalize their operations in China and India. Again, however, basic information is sparse on the degree to which private sector R&D in China and India is carried out by foreign firms as compared with indigenous firms. There are 800 R&D programs in China run by foreign companies and more than 200 in India, but how these numbers convert into shares of total private sector R&D expenditures, broken down by sector, is not reported.

Some general observations can be made, however, about the structure and content of private sector R&D and how they influence the course of events in China and India. Most importantly, on a global basis, more than 90 percent of private sector R&D is concentrated in the manufacturing sector. [30] The largest sectoral shares of global R&D, in rank order, are in six sectors: IT hardware and software, pharmaceuticals and biotechnology, automotive, electronics, chemicals, and aerospace. In this context, the Chinese economic strategy of manufacturing-led growth, with concentration in the IT and other advanced technology industries, has laid the foundation for the 19 percent annual growth in R&D, roughly the same growth as achieved for overall manufacturing in China. India, in contrast, aside from software, has had a relatively small, slow-growth manufacturing sector, hobbled by protection, lack of infrastructure, and overregulation, which is only belatedly being addressed, although with considerable success, by the National Strategy for Manufacturing.

MNCs have various reasons for establishing R&D programs in China and India. The two most important are lower cost and high quality of engineers and other technology-oriented personnel. Another reason is to have fully engaged production facilities in these large and growing markets, including R&D work targeted at product development within these markets. A key dimension of the structure of the R&D work, however, is the extent to which it is an integrated part of corporate global R&D programs, in effect a globalized supply chain management for R&D. And in this respect, there is a clear difference between China and India, which is becoming more pronounced as FDI grows more rapidly in India. In China, foreign firm R&D is more heavily directed to improved product development for the Chinese market, despite strong pressures by the Chinese government to attract more advanced R&D work. Foreign firms are concerned, especially in the context of the new independent innovation strategy, that advanced

[30] In 2005, 253 of the 300 largest corporate R&D programs globally were in the manufacturing sector, and they accounted for more than 90 percent of the $428 billion total expenditures. See the U.K. Department of Trade and Industry Annual Scoreboard (October 2006).

design R&D work in China will be either copied by Chinese competitors, including current joint venture partners, or stolen in violation of intellectual property rights. In India, in contrast, the government does not interfere in how foreign firms do their R&D work, and the less intense competition from Indian firms takes place within a more secure legal framework.

Three important examples illustrate the growing attraction of globally integrated R&D work in India by American firms. The GE Welch Technology Centre in Bangalore employs 2,500 engineers, engaged in both product design for the Indian market and more integrated, advanced design R&D work. IBM more than doubled its staff in India from 2004 to 2007, with engineers in R&D rising to 3,000, for everything from software to semiconductors to supercomputers. And in January 2007, Cisco announced that its engineering staff in India would be increased from 1,400 to 6,000 by 2010, at which time it would constitute 50 percent of global R&D. Moreover, as part of this bilateral integration initiative, 20 percent of the Cisco global leadership will be relocated to India, and manufacturing facilities will be built there.[31]

These large offshore R&D programs by MNCs for new product development contrast, up to this point, with R&D programs by indigenous Chinese and Indian firms, which have tended to be relatively smaller in size and far more concentrated on shorter term product application of existing technologies. This orientation is also beginning to change, however, and is discussed in greater detail in Chapter 4, in the context of Chinese and Indian firms evolving into competitive MNCs.

There is finally the qualitatively important role played by venture capitalists, stimulating technological innovation through the financing of small, R&D-intensive startup companies. Venture capitalists have moved aggressively into China and India since about 2000 and 2004, respectively, as discussed below.

Technology-Oriented Education

Technology-oriented education is the essential foundation for building advanced technology economies in China and India, and it is multifaceted. It supplies the human resources for R&D, industrial production, and related service industries. It is wide ranging, from granting doctoral degrees in engineering and science to core science and math education at the primary and secondary school levels. It is highly qualitative, not only in what subjects are taught, but in how they

[31] For the Cisco initiative, see *Business World*, January 15, 2007, pp. 24-28.

are taught. There are public and private sector roles in play. And there is political impact, as rapidly expanding education produces parallel growth in a more affluent and better-informed middle class.

A rapid, comprehensive expansion of technology-oriented education is under way in China and India, with important differences in form and content, but with a similar result of a very large scale output of graduates, comparable, in quantitative terms at least, to that of the United States and Europe. Unlike R&D and FDI, India has been more comparable to China in academic output, including some qualitative advantages, although overall China still has a substantial lead in education and is moving forward at a faster pace.

The presentation here starts with a broad look at university training, and engineering graduates in particular, highlighting relative strengths and weaknesses between China and India. This is followed by briefer discussion of several other areas of education, with a focus on those in greatest need of improvement. The section concludes with commentary on what are called "three burgeoning forces" within the educational framework.

University Training

China has invested heavily in university education, with the spectacular result that the number of college graduates quadrupled in ten years, from 1995 to 2005, from 800,000 to over 3 million. This compares with a relatively stable 2.4 million graduates in the United States. The large Indian system, with over 300 universities, dates back to the early years of independence, did not suffer a severe setback as did China during the Cultural Revolution of the 1970s, and graduates about 2.5 million students per year. Thus, China and India are turning out roughly the same number of university graduates as the United States and the EU, and more than double the level of Japan.

There are major weaknesses in Chinese and Indian education, however, compared with the advanced industrialized grouping, particularly concerning whether college graduates are qualified for jobs in the modern industrial sector. Assessments of relative weaknesses vary and are largely episodic. Consequently, only general comments are offered here. China continues to devote major resources to raise the quality of education in the recently expanded university system, through public funding and the application of tuition payments up to 40 percent of costs. Relationships with U.S. and other foreign universities, joint programs with private companies for technical training and research, and higher pay for "star performer" professors all support the raising of Chinese performance standards.

The Indian university system, in contrast, suffers from severe budget constraints, a general prohibition on tuition payments at public universities, and low salary caps for faculty, which is causing an exodus of the best professors to higher paying private sector jobs. Wide-ranging concern is expressed about the low level, if not decline, of academic standards in many universities, but there is not yet a clearly stated, adequately financed government response to the problem. Another difficult issue is political pressure to broaden the use of lower admission standards for Scheduled Castes/Tribes and Other Backward Classes, which would further weaken Indian academic standards. Private sector collaboration has also been more restrictive and less attractive in India than in China, but this is beginning to change. Most of the large Indian software and business services companies now have large in-house training programs of at least several months to provide the technology-oriented skills necessary for productive employment.

Engineering graduates have received the most attention in view of their central role in the development of advanced technology industry, and again the Chinese growth performance is the more impressive. The number of engineering undergraduates roughly doubled in China from 1995 to 2005. Table 2-4 shows the numbers of engineering, computer science, and information technology degrees granted in 2004 by the United States, China, and India, broken down by four-year bachelor and two- to three-year subbaccalaureate degrees. In both categories, China has two to three times more graduates than either the United States or India. India is on a par with the United States for total graduates

Table 2-4

Number of Bachelor and Subbaccalaureate Degrees in Engineering, Computer Science, and IT, 2004

	United States	**China**	**India**
Bachelor Degrees	137,437	351,537	112,000
In Engineering	52,520	NA	17,000
Computer Science, Electrical, and IT	84,917	NA	95,000
Subbaccalaureate Degrees	84,898	292,569	103,000
In Engineering	39,652	NA	57,000
Computer Science, Electrical, and IT	45,246	NA	46,000
Total Number of Degrees	**222,335**	**644,106**	**215,000**

Source: Gary Gereffi and Vivek Wadhwa, *Framing the Engineering Outsourcing Debate: Placing the United States on a Level Playing Field with China and India*, Duke University, December 2005, p. 5.

but has a lower share of four-year bachelor degrees, particularly in engineering.

The National Science Foundation (NSF) pays special attention to doctoral degrees, as most directly relevant to technological innovation and advanced technology industry development. Table 2-5 presents the number of engineering doctorates granted, from 1995 to 2003, in China, India, the United States, the EU, and Japan. In this case, in 2003, the EU had the highest number of doctorates by far, at 11,263, followed by China, with 6,573, the United States, with 5,265, and Japan, with 3,921. India lagged distantly behind, with only 779 doctorates.

Table 2-5

Engineering Doctoral Degrees

	1995	**2000**	**2003**
European Union (15)	9,501*	8,946	11,263
China	1,659	4,484	6,573
United States	6,008	5,320	5,265
Japan	2,791	3,800	3,921
India	335	723	779

*1996.
Source: NSF Science and Engineering Indicators 2006.

Qualitative differences in engineering graduates have received widespread attention, related to the globalization of R&D and export competitiveness. Both China and India have top-rated schools. India has had a qualitative edge in this regard, with its highly acclaimed seven Indian Institutes of Technology (IITs) dating back to 1951, which enroll 15,000 undergraduate and 12,000 graduate students each year. In December 2006, the creation of three additional IITs received final government approval, and a few more are planned. Selected Chinese universities, such as Tsinghua and Peking, also have high quality reputations. Most engineering schools in both countries, however, are in sore need of improvement. The general complaint in China is that engineering graduates, although well rounded in theory and core curriculum, lack initiative and practical skills for business and require years of experience and training in private companies to become fully productive. Many foreign companies have nevertheless been willing to make this human resource investment, with positive results. The problem in India reflects the low quality of faculty and the lack of laboratory and other facilities at Indian universities. At the same time, however, Indian graduates have the advantages of a more probing and market-oriented mind-set and fluency in English. Many companies selectively recruit engineers down to the second- and third-tier universities with favorable results, although supplemental in-house training, as noted above, is often required.

An additional component of engineering and science recruitment in both countries, of critical qualitative importance, is the return of overseas diaspora graduates, with degrees and often business experience abroad, principally in the United States and Europe. This subject is discussed in the final section of Chapter 4, "The Amazing Diaspora Connection."

Other Educational Priorities

The first priority for education, of course, is to have strong primary and secondary schooling available to all, or almost all, of the population. China and India both have major weaknesses in this area, although once again China has advanced further toward the goal. The literacy rate in China is 95 percent for men and 87 percent for women, compared with 73 percent and 48 percent, respectively, in India. The biggest gaps are in rural areas, where qualified teachers and physical accommodations can be severely lacking, while the provincial governments in these areas are the poorest.

Another area of education relevant to the development of advanced technology industry is business school education, and the granting of MBAs in particular, which is a high priority for both countries. India again was off to an earlier start, through its six Indian Institutes of Management (IIMs), dating from 1961, where 1,200 students are selected each year through intense competition to pursue a two-year MBA program. The first 86 MBAs in China were awarded only in 1991, but MBA programs have since proliferated, and in 2006 230 MBA programs, including 60 to 70 joint programs with U.S. and other foreign business schools, graduated 20,000 students. Chinese students, including upwardly mobile managers, are willing to pay the cost, about $25,000 for two years at a top-rated school such as Tsinghua University, to obtain an average 80 percent pay increase upon graduation. Many of the newly created Chinese business schools are criticized, however, as of low quality, with graduates lacking in both confidence and the spirit of risk-taking, and being unable to express themselves.[32]

One crucial dimension of education for globalized industry and services is English language capability. Here India has a big advantage over China, with English the national language, including for all official documents. English is not generally spoken in the Indian countryside, however, and fluency in English remains limited to a

---

[32] *Wall Street Journal*, September 20, 2005; *BusinessWeek*, November 4, 2006, p. 106.

minority of the population. Some states are even moving away from English toward primacy for local languages,[33] although fluency in English is understood by all to be a prerequisite for advanced or university-level education. China has designated English as a compulsory subject down to the primary school level, and some university training, including at business schools, requires English capability. China is thus actively seeking to level the playing field with India with respect to English-language capability, but remains far behind.

Three Burgeoning Forces Within the Educational Framework

All of the foregoing specific components of the education systems in China and India come together in a highly dynamic setting of forward movement and transformation. The driving forces, moreover, are not limited to the education ministries, or even the top levels of government. They also include other powerful forces within the two emerging knowledge-based societies, with education at the center. Three forces burgeoning forth and interacting are a market-oriented economic setting, a growing private sector role, and political momentum for quality education open to all.

A market-oriented economic setting.—With 8 to 11 percent annual growth from rapid transformation to a knowledge-based economy, the rewards for a strong, skill-based education are quick and bountiful, and parents as well as students are willing to work and pay the price to get it. The rise from poverty to middle class living for tens of millions of young people each year is omnipresent, from direct observation to TV stories to Internet blogs. The deep cultural commitment to family in both countries, moreover, puts top priority on education for children, even in the most backward classes. In China, the limit of one or two children per family intensifies the focus on providing the best education for those one or two, no matter what the cost. And if public schools are inadequate, parents are willing to pay for improvements in public schooling or for alternative superior education in private schools. For advanced education, a similar market exists, with working people willing to pay the price for clearly recognized large rewards. In effect,

---

[33] A striking example of this retrogressive policy is the state of Karnataka, with high-tech Bangalore its capital. The state government decreed that all new private schools must teach in Kannada, the local language, and threatened to close down 2,000 private schools that teach in English. This was a joint initiative by traditional, anti-imperialist politicians and the state school system, whose "rotten" performance and lack of qualified English teachers have resulted in a large exodus of students to English-speaking private schools. See *The Economist*, November 11, 2006, p. 50.

market demand for quality education is surging, the supply needs to catch up, and a rising market price is making this happen.

A growing private sector role.—Private sector education is growing rapidly in both China and India, although hard analysis is scarce, in large part because the official education establishments resist recognizing and accrediting such competition. Private schools are flourishing in both countries, however, with abundant anecdotal reports. For India, the NIIT initiative was recounted earlier, and at the primary and secondary school levels there are both traditional elite schools and low-budget schools for the lower classes. Private companies are also playing a more active role, with financial support for education up through university-level training, linked to job offerings. China's private schools and corporate training programs have been much larger, and continue to grow. The products of the information technology sector themselves add potential for more productive education, including the training of teachers as well as students in remote regions through software products used over the Internet, which private companies and schools are more agile in providing than public sector bureaucracies. For both governments, a laissez-faire policy, by design or neglect, is unleashing a momentum for education based on the private sector offering a superior product to parents and students eager to pay the price.

Political momentum for quality education open to all.—The ever more broadly based demand for quality education is having a political impact, as a better-informed population makes its priorities known, and becomes more willing to act to achieve them. Political pressures are building for a more responsive public sector education system and a more open private sector component. The educational establishments, especially in India, are feeling growing public pressures for reform. The political scenario for wide-ranging reform in response to rapid growth in middle-class status and aspirations will be very different between authoritarian China and democratic India, a complex subject addressed in Chapter 5. But there is no doubt that the political pressures for educational reform and improvement will continue to grow in both countries.

Foreign Direct Investment (FDI)

A large inflow of FDI played a decisive role in the development of advanced technology industry in China from 1995 through 2005, during which time the inflow was far smaller in India. This is now changing, however, with FDI rising rapidly in India while leveling off and perhaps beginning to decline in China in the technology-intensive manufacturing and related services sectors. The FDI flow into India, in any event, had long been understated because official Indian statistics, unlike Chinese statistics and international practice generally, did not count reinvested earnings. India recently began providing figures for total FDI including reinvested earnings, backdated to 2000, and the inclusion results in a more than 70 percent increase during the period 2000-2005, and smaller, yet still significant increases for 2006 and 2007.

Table 2-6 presents FDI inflows for China and India, on a comparable basis, including reinvested earnings, for 2000-2006. In 2005, FDI into China reached a new high of $72.4 billion and then declined slightly to $69.5 billion in 2006. These figures included, for the first time, large investments in banking and other financial services, at $12 billion and $13 billion in 2005 and 2006. FDI in manufacturing correspondingly declined from a high of $43 billion in 2004 to $42 billion in 2005 and to $40 billion in 2006, or from 70 percent of the total in 2004 to 58 percent in 2006. For India, FDI rose by 42 percent in 2004 to $6.1 billion, by 13 percent in 2005 to $7.7 billion, and then surged by 153 percent in 2006 to $19.5 billion. The government projects FDI for the financial year through March 2008 to be substantially higher, perhaps as high as $30 billion. The net result is that the China/India ratio of FDI, long reported in favor of China by 10 to 20 to 1 has now dropped sharply to about 3 to 1, and this ratio is likely to decline further.

Table 2-6

Foreign Direct Investment in China and India

(Disbursements, $ billions)

	China	**India***
2000	40.5	4.0
2001	46.9	6.1
2002	52.7	5.0
2003	53.5	4.3
2004	60.6	6.1
2005	72.4	7.7
2006	69.5	19.5

*Financial year, April through March of the following year; includes reinvested earnings.
Sources: Fact Sheet on Foreign Direct Investment, Department of Industrial Policy and Promotion, Ministry of Commerce and Industry, Government of India; and for China, www.fdi.gov.cn/pub/fdi.

There are also major differences between China and India in the sources of FDI, although precise figures are not available because of large indirect flows through third countries and "free ports," as shown in Table 2-7. For China, FDI from Asia dominates, with 58 percent of the listed sources, or two and a half times that of the United States and the EU combined. The $2.1 billion from Taiwan, moreover, greatly understates the actual flow because most Taiwanese investment enters indirectly via Hong Kong and the free ports. For India, in contrast, almost five times as much investment was recorded from the United States and the EU as from the five listed Asian sources—$3.7 billion versus $0.8 billion in 2006—but these figures pale by comparison with the $6.4 billion from Mauritius.

Table 2-7

Foreign Direct Investment in China and India by Source, 2006*

	China		**India****	
	$ billions	**Percent**	**$ billions**	**Percent**
Total	69.5	100	15.7	100
United States	6.6	10	0.9	6
EU	6.7	10	2.8	18
Subtotal	13.3	20	3.7	24
Japan	4.6	7	0.1	1
South Korea	3.9	6	0.1	1
Singapore	2.4	4	0.6	4
Taiwan	2.1	3	Neg	
Hong Kong	20.2	29	Neg	
Subtotal	33.2	48	0.8	5
Virgin Islands	11.2	16	Neg	
Mauritius	Neg		6.4	41
Other	11.8	17	4.8	31

* For India, the financial year April 2006-March 2007.
**Does not include reinvested earnings.
Sources: For India, Fact Sheet on Foreign Direct Investment, Department of Industrial Policy and Promotion, Ministry of Commerce and Industry, Government of India; and for China, www.fdi.gov.cn/pub/fdi.

By sector, FDI has been more diverse in India than in China, where manufacturing, until recently, dominated. For 2002-2006, Indian investment by principal sector was 23 percent for the electronics sector, including IT hardware and software, 13 percent for telecommunications, 12 percent for financial and nonfinancial services, and 10 percent for chemicals, including pharmaceuticals.

These are the broad characteristics distinguishing FDI in China and India. The role of FDI in the development of advanced technology industry, in particular, is now discussed for each country, with commentary on the outlook ahead. FDI outflows, although smaller than inflows, have also been rising for both countries, but this subject is left for Chapter 4, related to the emergence of Chinese and Indian multinational corporations.

FDI as the Decisive Catalyst in China

The rapid rise of FDI in China, beginning in the mid-1990s, was described earlier and is shown in Figure 2-1. The central objective since 1995 has been the development of advanced technology industries, and the IT sector most of all. FDI played a central role in terms of the growth in R&D and, most importantly, for growth in exports. The full story of how FDI in China became the decisive catalyst for the development of Chinese advanced technology industry has been recounted elsewhere,[34] and the presentation here is limited to three key relationships that have driven the process and are now each undergoing significant change: Taiwan as number one and the most deeply integrated foreign investor; the U.S. and European longer term strategies; and the fading East Asian export platform issue.

Taiwan: Number one and the most deeply integrated investor.— There is general agreement that Taiwan is the number-one investor in China, but there are no reliable figures as to precisely how much investment goes in because most of it enters indirectly from Hong Kong, the free ports, and other third countries and is thus not recorded as coming from Taiwan. Official Chinese figures show only $2.2 billion of investment from Taiwan in 2005, while observers have estimated much higher levels, ranging from $10 billion to $30 billion. Taiwanese investments are much smaller, on average, than American or other foreign investments, and some projects are split into two or more separate investments so as to avoid stringent Taiwanese restrictions, which generally apply to investments of $100 million or more. The common Chinese ethnicity and language enable Taiwanese firms to blend into the Chinese industrial community, often involving cross-border family ties.

Taiwanese investment has played a major role in creating a modern manufacturing industry in China, particularly in the IT sector. On the ground Taiwanese managers and engineers understand that Chinese employees need private sector learning experience, and they are willing to provide it. A manager of Taiwanese PC maker MiTAC International Corp. explains: "We set up an R&D center in Shanghai in 2000, in which Chinese engineers work on the designs of our mainboards, PCs, servers and information appliances. . . . Basically all tasks could be

---

[34] See Preeg, *op. cit.*, 2005, "The Decisive Catalyst: Foreign Direct Investment," pp. 33-60.

done in China with the support of our Taiwan R&D team. . . . We teach them how to work in a market-oriented company where time-to-market is a very critical factor."[35]

A comprehensive assessment of Taiwanese/Chinese integration in the IT sector by the Rand National Defense Research Institute (NDRI) concluded: "The scope and scale of trade and investment flows across the Taiwan Strait have increased dramatically in recent years, driven in large part by the increasing integration of the information technology (IT) sectors. . . . Taiwanese-invested companies produced more than 70 percent of the electronics made in China. . . . The share of Taiwan's IT hardware production in mainland China reached 60 percent in 2003."[36]

The deepest Taiwanese-Chinese, or "Greater China," concentration is in the semiconductor sector, where the two together run second to the United States in terms of production and are closing the gap. Taiwan dominates the global foundry industry for semiconductors, including the two largest companies, Taiwan Semiconductor Manufacturing Co. (TSMC) and United Microelectronics Corp. In 2004, for the first time, the Taiwanese government approved the construction of a semiconductor factory in China by TSMC, and other Taiwanese companies are pressing for approval to follow in. The largest Chinese company, Semiconductor Manufacturing International Company (SMIC), founded in 2000 by Taiwanese-born and U.S.-trained Richard Chang, began production in 2004 of advanced design, 12 inch semiconductors. For semiconductor design, which is critical to technology innovation, 14 of the largest firms are American, followed by 5 Taiwanese and 1 Canadian. In China, with strong government incentives, there are now about 400 chip design firms, most only a few years old, one-third of which are foreign companies, including Taiwanese.

Taiwanese investment in China extends beyond the IT sector and continues to broaden and deepen, with political as well as economic implications. China is now Taiwan's number-one trading partner, with the United States a steadily receding number two. The Taiwanese business community is thus more and more heavily dependent on a positive, open trade relationship with China and opposes conflict related to Taiwanese independence. An important issue is the on-again, off-again discussion of an agreement on the "three links"—direct trade, transport, and postal service—between Taiwan and China. Taiwan

---

[35] "China Special Firms Shift R&D to China," http://www.neasia.nikkeibp.com/nea/200204/srep_178979.html. The quotation from the Taiwanese manager is from this report.

[36] Michael S. Chase, Kevin L. Pollpeter, and James C. Mulvenon, *Shanghaied? The Economic and Political Implications of the Flow of Information Technology and Investment Across the Taiwan Strait* (Rand National Defense Research Institute, July 2004), p. xiii.

business leaders press strongly for an agreement, with Y. C. Wang, chairman of Formosa Plastics, warning that if the three links are not opened, "all our businesses in China will lose their competitiveness in three years."[37] The *Taiwanese News* countered with the political argument that the three links would "totally undermine Taiwan's national security . . . leaving it fewer cards to play in its attempt to resist China's ever-mounting pressure and blackmail."[38] Where the China/Taiwan political relationship is heading over the longer term cannot be predicted, but it is clear that the economic component of the equation, anchored in Taiwanese investment on the mainland, is growing in relative strength and is a deepening force for political reconciliation and a more collaborative Greater China in international economic relations.

The United States and Europe: A longer term strategy.—The United States and the EU together account for only about 15 to 20 percent of FDI into China, depending on how much comes in indirectly from the free ports or elsewhere. In 2004, officially recorded investment was in rough parity, $3.9 billion U.S. and $4.2 billion EU. In 2005, however, EU investment rose to $5.2 billion, while U.S. investment declined to $3.1 billion, and in 2006 EU investment was $5.3 billion compared with $2.9 billion by the United States.

U.S. and European investment is more widely dispersed by sector compared with the concentration in manufacturing by the East Asian investment from Taiwan, South Korea, and Japan, and in general is more heavily targeted on the domestic Chinese market. U.S. and European firms have higher visibility than most East Asian counterparts and are widely recognized for establishing highly valued R&D programs and producing more advanced design products. They are also more likely to have a more positive corporate image, in terms of higher labor and environmental standards and greater emphasis on training and promotion based on merit for their employees.[39]

U.S. and European investment through 2004 was heavily in manufacturing, although less so than by the East Asians. Forty-nine percent of U.S. investment during 2003-2006 was in manufacturing, of which 22 percent was in chemicals, including pharmaceuticals, 17 percent in transportation, 14 percent in computers and electronic products, 5 percent in electrical equipment and appliances, and 5

---

[37] *Ibid*, p. 14.
[38] *Ibid*, p. 16.
[39] See Ernest H. Preeg, *U.S. Manufacturing Industry's Impact on Ethical, Labor and Environmental Standards in Developing Countries: A Survey of Current Practices* (MAPI 2001). This survey contains more than 300 examples of specific "good practices," including 18 by U.S. companies in China.

percent in machinery. Other prominent sectors were wholesaling, 11 percent, and mining, 10 percent.[40]

Another important shift under way in the orientation of FDI, with the Americans and the Europeans out front, is toward the central and western provinces and away from the high concentration in the coastal industrial zones. In 2004, the Chinese government placed urgent priority on job-creating investment in the inner provinces, accompanied by large expenditures on infrastructure and financial incentives for investors. Wages, real estate, and other costs are lower than in the coastal cities. American and European firms have been quick to respond. Emory Williams, chairman of the American Chamber of Commerce in Beijing, stated that, by late 2006, 50 percent of U.S. member companies had already invested in second-tier cities.[41] In 2003, Motorola established an R&D center in Chengdu that the local government helped finance, and Alcatel, Ericsson, Nokia, and Microsoft quickly followed in. In addition to offering lower costs, the local market in the western provinces is growing rapidly. Intel Corporation justified $450 million of investment for two plants in Chengdu largely because computer sales in the region were growing by 45 percent per year.[42]

Another noteworthy development is the arrival in force of American venture capitalists, who, as in Silicon Valley and elsewhere, provide the start-up financing for smaller firms with promising technological innovation. Patrick McGovern was the first American venture capitalist in China in 1992, and he built his IDC Technology Investment Inc. investment firm from an initial $1.7 million to $37 million in 2005, for small, largely high-tech investments averaging just over $1 million.[43] It was not until 2000, however, that foreign venture capitalists began to arrive in large numbers, and by 2004 there were 253 financings in China totaling $1.3 billion, 80 percent by foreign venture capitalists.[44] The venture capital market in China is a truly Wild West (or East) setting, within which it is playing an important role in nurturing innovative start-up and other small companies.

East Asia: The fading export platform issue.—Taiwan, South Korea, and Japan, the top three investors in China, have been overwhelmingly oriented to the manufacturing sector, on the order of 90 percent of total

---

[40] U.S. Department of Commerce, *Survey of Current Business*, September 2007. The percentages are limited to capital outflows and do not include reinvested earnings or financing from banks in third countries.

[41] "Go West, Westerners," *BusinessWeek*, November 14, 2006, pp. 60-61.

[42] *Ibid.*

[43] *BusinessWeek*, January 24, 2005.

[44] *Zero2IPO-China Venture Capital Annual Report 2004*; and Pieter Bottelier, "Venture Capital and Innovation in China," presented at SCID Conference on "China's Policy Reforms: Progress and Challenges" (Palo Alto, October 14-16, 2004).

investment. Much of this production, particularly in the IT and consumer electronics sectors, involves imported components incorporated in final products for export. Such cross-border supply chain production is common by multinational corporations, but when the share of imported content reaches high levels, the term "export platform" is used to indicate that the exporting country is merely a platform for assembling imported components for reexport. This export platform issue has received prominent attention with regard to Chinese exports, particularly by East Asian companies.

There has been substantial "export platforming" by the East Asians in recent years, which has resulted in very rapid growth in exports of components by the East Asians to China, largely for reexport to the United States and Europe. One result is larger trade surpluses with China by the East Asians and larger U.S. and European bilateral trade deficits with China. The share of import content in exports, however, has not been nearly as large as often reported, as explained in Chapter 3, where updated estimates for the IT and electronics sectors are provided. Moreover, whatever the domestic Chinese share of content was a few years ago, it is considerably higher today, and continues to rise. Foreign investors steadily increase Chinese value added, both in their internal operations and through developing supplier relationships with Chinese firms, including through training and technical support to improve the quality standards of in-country suppliers. In effect, a progressively larger share of the components arriving in the export processing zones is of Chinese origin. The net result is a progressively higher share of Chinese value added in exports, and a declining trade effect from "export platforming."

* * *

In conclusion, FDI has been the decisive catalyst for the development of advanced technology industry in China, but it is now in transition in important ways, which is changing its character and impact. By sector, there is the relative and possibly absolute decline of FDI in the manufacturing sector, while a growing share moves into financial, distribution, and other service industries. FDI in financial services can produce large, multi-billion-dollar investment figures, but it does not create the high-tech jobs and R&D programs as does FDI in manufactures. A shift toward investment for the domestic market, including in the inner provinces, rather than for exports, also involves a different industrial mix, with greater emphasis on such products as construction and environmental equipment, household goods, and consumer products tailored to the Chinese consumer. The fading export platform issue, in terms of progressively higher Chinese value added in the IT and electronics sectors in particular, will reduce East Asian export growth to China, and diminish if not end East Asian trade surpluses with China.

These various shifts in the pattern of FDI, moreover, will take place within the broader context of the anticipated structural adjustment in the Chinese economy, away from export- to more domestically oriented growth, driven largely by a major revaluation of the yuan, which is a central topic for China in Chapters 5 and 8. This will involve a major squeeze on the manufacturing sector, already faced with overcapacity in a number of industries. In these difficult circumstances, increased economic nationalism and preferential treatment for Chinese firms within the independent innovation strategy will further discourage FDI in manufacturing.

This difficult adjustment does not mean a collapse of FDI in China. Most companies are in China for the long haul of high growth in a China that will become considerably more affluent in dollar terms as the yuan rises. And there is ample growth opportunity for foreign and Chinese companies as long as Chinese economic nationalism does not become excessive. But 2005-2006 could have been the FDI peak for China, especially in the manufacturing sector.

FDI as an Emerging Catalyst in India

FDI in India was inconsequential through the 1990s, then rose substantially beginning about 2000, and has surged dramatically since 2004. The central questions for Indian growth performance now and in the years ahead are why this rapid growth in FDI emerged, and will it continue or even accelerate further. The assessment here is that the higher level of FDI will continue and likely accelerate, contingent on further reform measures by the Indian government, discussed in Chapter 5. The initial presentation here is limited to the basic facts about the recent surge in FDI in terms of geographic sources, sectoral distribution, and location within India, plus a commentary on the unclear policy framework within which this is all happening.

FDI in India, even after adjusting to include reinvested earnings, averaged only about $3 billion per year in 1991-1999, or less than one-tenth of that in China, and then, as shown in Table 2-6, rose to the range of $4 billion to $6 billion per year in 2000-2004. The big surge, however, only began in 2005, when FDI rose from $6.1 billion in 2004 to $7.7 billion in 2005 and to $19.5 billion in 2006.

The sources of this rapidly rising inflow are presented in Table 2-8 for 2003-2006, although unfortunately these figures, like those in the subsequent tables on distribution by sector and state, do not include reinvested earnings. For example, in 2006, the total was $15.7 billion rather than the $19.5 billion including reinvested earnings. The table lists the top ten cumulative national investors since 1991. Mauritius stands out by far as the number-one source, with its share of the total increasing from 22 percent in 2003 to 40 percent, or $6.4 billion, in 2006. This rise is the result of the India-Mauritius double taxation avoidance agreement, which provides an attractive tax shelter for investors in India but, at the same time, clouds the picture of the national origin of investor companies.

Table 2-8

Foreign Direct Investment in India by Country Source

($ millions)*

	2003	**2004**	**2005**	**2006**
Total	2,634	3,754	5,546	15,726
Mauritius	**567**	**1,129**	**2,570**	**6,363**
United States	**360**	**669**	**502**	**856**
Europe	**818**	**707**	**759**	**2,815**
Germany	81	145	303	120
United Kingdom	167	101	266	1,878
Switzerland	45	77	96	56
Netherlands	489	267	76	644
France	36	117	18	117
East Asia	**89**	**345**	**543**	**734**
Singapore	37	184	275	578
Japan	28	126	208	85
South Korea	24	35	60	71

*By financial year, from April of the listed year through March of the following year; does not include reinvested earnings.

Source: Fact Sheet on Foreign Direct Investment, Department of Industrial Policy and Promotion, Ministry of Commerce and Industry, Government of India.

Among the other nine countries listed, the United States and the five listed European countries have long dominated, with investment several times larger than that of the three listed Asian investors through 2006, although investment from the East Asians has been rising at a faster pace. The U.S.-European relationship, moreover, has long favored Europe, with the five listed Europeans investing more than twice as much as the United States during the 2003-2006 period.

The East Asian growth in FDI since 2003 has been shared by Singapore, Japan, and South Korea, with Singapore rapidly pulling ahead of the others, up to $578 million in 2006. This increase by Singapore is principally the result of the June 2005 bilateral Comprehensive Economic Cooperation Agreement (CECA), which goes well beyond a free trade agreement, covering services, investment, customs cooperation—a big issue in India—science and technology, education, and intellectual property. Investment from China and Hong Kong remains relatively small and India carefully screens this investment, largely on national security grounds, which has delayed if not denied approval of investment proposals in the telecommunications, airport equipment, and port operations sectors. Chinese investment is rising, however, and will probably accelerate.

Table 2-9 presents FDI in India by principal sector, showing considerable diversity, with financial and nonfinancial services in first place, at 22 percent of the total during 2003-2006, followed by electrical equipment at 20 percent, and telecommunications and transportation industry at 5 percent each.

Table 2-9

Foreign Direct Investment in India, Ten Top Investors by Sector*

($ millions)

	2003	**2004**	**2005**	**2006**	**Percent 2003-2006**
Total	2,634	3,754	5,546	15,726	100
Financial and Nonfinancial Services	269	469	581	4,749	22
Electrical Equipment (Including Computer Software and Electronics)	532	721	1,451	2,733	20
Telecommunications	116	129	680	521	5
Construction Activities	---	152	151	985	5
Transportation Industry	308	179	222	466	4
Drugs and Pharmaceuticals	109	292	172	215	3
Chemicals	20	198	447	206	3
Fuels (Power and Oil Refinery)	113	166	94	250	2
Housing & Real Estate	---	0	38	467	2
Food-Processing Industries	111	38	42	98	1
Other	1,056	1,410	1,668	5,036	33

*By financial year, through March of the following year; does not include reinvested earnings.
Source: Fact Sheet on Foreign Direct Investment, Department of Industrial Policy and Promotion, Ministry of Commerce and Industry, Government of India.

Table 2-10 presents FDI by principal city and state over the period January 2000-June 2007, which is striking by its disparity. Mumbai and New Delhi, the two “Tier-I mature destinations,” dominate with 67 percent of the total over the seven and a half year period, while “Tier-II transition destinations” Chennai and Bangalore account for an additional 19 percent. Near the bottom is Kolkata, formerly Calcutta, with only 1 percent, which has historical irony in that for three centuries Calcutta, as home base for the East India Trading Company, was the commercial center of India. This disparity of investment is changing, however, as the “Tier-III emerging destinations” offer lower costs, improved infrastructure, and more attractive fiscal and other incentives to lure job-creating foreign investments.[45] Even Kolkata and Kerala, governed by communist and other anti-capitalist political parties long opposed to foreign investment, are now actively seeking and attracting foreign investors, as recounted earlier related to new SEZs.

Table 2-10

Foreign Direct Investment In India by City and State*

(January 2000-June 2007)

Principal City**	States Covered	Amount ($ Millions)	Percent of Total
Mumbai	Maharashtra, Dara and Nagar, Haveli, Daman and Diu	9,310	34
New Delhi	Part of Uttar Pradesh and Haryana	9,188	33
Chennai	Tamil Nadu and Pondicherry	2,728	10
Bangalore	Karnataka	2,635	9
Hyderabad	Andhra Pradesh	1,406	5
Ahmedabad	Gujarat	1,092	4
Chandigarh	Punjab, Haryana, Himachal Pradesh	364	1
Kolkata	West Bengal, Sikkim, Andaman & Nicobar Islands	357	1
Panaji	Goa	183	1
All other cities/states		499	2
Total ***		**38,124**	**100**

*Does not include reinvested earnings.
**The regional offices of the Reserve Bank of India, which provides the information.
***An additional $10.5 billion of inflow is in categories not broken down by city and state, and not included in the table.
Source: Fact Sheet on Foreign Direct Investment, Department of Industrial Policy and Promotion, Ministry of Commerce and Industry, Government of India.

The policy framework for stimulating the surge in FDI beginning in 2004 is complex and evolving. In the largely decentralized Indian political system, competition among states to attract job-creating foreign investments is following in the earlier Chinese path, but local politics and the ever-present license Raj can slow or stifle the movement toward greater openness to investment. National policy is drawn heavily from the August 2002 Report of the Steering Group on Foreign Direct Investment, sponsored by the Government Planning Commission. This comprehensive 119-page document by a distinguished 12-member group documented the relatively poor performance of India in attracting FDI during 1989-2001 and recommended a number of policy changes to stimulate a rapid increase over the ensuing five years. The policy

[45] See "Tier-III Cities Set for Rapid Growth," *Business Standard*, November 24, 2006.

recommendations include higher foreign equity caps and a shift from review of new investments by the highly bureaucratic Foreign Investment Promotion Board to automatic approval. This recommendation would bring the entire manufacturing sector, except defense industry, to a 100 percent equity limit via the automatic route. Similar although less far-reaching liberalizing steps were recommended for the mining, airport, oil and gas pipeline, telecommunications, and financial services sectors. Much of this has now been implemented. A special appeal was made to liberalize the 2000 SEZ law which was "still not fully on par with the Export Zones of China," and this was implemented in February 2006. Highlighted throughout the report, however, are warnings about the hurdles still to be faced by both domestic and foreign investors, at the federal and state levels. This remains the current mixed state of policy but with the political as well as the economic policy momentum supporting the strong upward trend in FDI.

A final comment about FDI in India, as in China, concerns the emerging venture capital market. The first Silicon Valley venture capitalist in India was Bill Draper in 1995, but his $55 million fund, Draper International, closed in 2000 after only one Indian company had been financed, and the venture capital market in India remained generally unattractive through 2004.[46] But like FDI in general, venture capital took off in 2005 and 2006, with over 44 U.S.-based VC firms in India by 2006, for an average capital of $100 million. The number of VC and PE (private equity) investments in India rose from 71 in 2004 to 146 in 2005 to an estimated 311 in 2006.[47] Moreover, this investment growth is diversified, with the IT share of the total dropping from 66 percent in 2000 to 23 percent in 2006, offset by rising shares of 19 percent for other manufactures, 10 percent for financial services, and 8 percent for medical and health care. In short, the entrepreneurial start-up investors, which have driven so much of U.S. advanced technology industry innovation over the past decades, are rapidly coming to life in India.

---

[46] See AnnaLee Saxenian, "International Mobility of Engineers and the Rise of Entrepreneurship in the Periphery" (Research Paper No. 2006/142, United Nations University, November 2006), pp. 17-19.

[47] See Alok Aggarwal, "Is the Venture Capital Market in India Getting Overheated?" (August 21, 2006), at Alok.Aggarwal@evalueserve.com. He warns of at least a temporary glut from the rapid inflow of venture capital and offers eight "best practices" for venture capitalists in India.

Chapter 3

TRADE AND EXPORT COMPETITIVENESS IN ADVANCED TECHNOLOGY INDUSTRIES

International trade data are the most revealing for the rise of China and India toward becoming advanced technology superstates. Merchandise trade statistics are the most current economic indicator—a 1 month time lag for merchandise trade for China, 2 months for the United States, and about 6 to 12 months for India—and are the most detailed by sector and product. Trade figures are also the direct measure of export competitiveness, which largely defines relative economic performance in the manufacturing sector and, to an increasing extent, in technology-oriented service industries.

Current and detailed figures are especially important at this time for China and India because of the rapid growth in the volume and changes in composition and geographic orientation of trade. In 2000, U.S. exports of manufactures were almost three times that of Chinese exports—$624 billion versus $224 billion, while only six years later, in 2006, China surpassed the United States, with $916 billion compared to $786 billion by the United States. Indian exports of information technology (IT) and other business services increased from $31 billion in 2004 to $60 billion in 2006, an almost doubling in two years.

The presentation here begins with a comparative presentation of Chinese and Indian trade, by sector and geographic orientation. This is followed by a closer look at Chinese trade in advanced technology industries and in Indian exports of software and other business services, focal points of the rapid advance up the technology ladder in the exports of the two countries. The final section examines the U.S. bilateral trade relationships with China and India, characterized by high growth and growing U.S. deficits.

Chinese and Indian Trade: The Broad Picture

Table 3-1 presents Chinese and Indian merchandise trade by broad sector, from 2000 to 2006. Four relationships stand out. First, Chinese trade far exceeds that of India, eight times larger for exports in 2006, and four times larger for imports. Second, both countries experienced extraordinary growth in trade from 2000 to 2006, up by 289 percent for Chinese exports and 252 percent for imports, while the corresponding Indian growth was 183 percent and 268 percent. The growth in Indian trade is understated, however, because of shortcomings in official

Indian trade statistics, whereby "provisional" figures, which are released about two months after the end of the year, and appear in Table 3-1, are much lower than the final figures that are delayed by almost a year. This understatement also obscures an important recent trade development, namely that Indian trade probably grew faster than Chinese trade in 2006, by over 30 percent for exports and imports, based on comparable provisional levels of trade in 2005 and 2006, compared with Chinese export and import growth of 27 percent and 20 percent. Confirmation of this higher growth in Indian trade in 2006 awaits the release of the final trade figures in early 2008. A plea from the author is consequently that the government of India, as it aspires to build a modern industrialized economy, do a better job of compiling its trade statistics, perhaps even moving to a calendar year as do all other major trading nations.

The third noteworthy relationship is that the manufacturing sector dominates merchandise trade throughout. In 2006, manufactures accounted for 95 percent of Chinese exports and 76 percent of imports. Chinese imports of mineral fuels, principally petroleum, and industrial raw materials, in contrast, together amounted to only 22 percent of total imports, despite widespread reporting about rapid growth and high oil and other commodity prices. For India, manufactures accounted for 67 percent of exports and 59 percent of imports in 2006, with a much higher, 37 percent share of imports for mineral fuels and industrial raw materials, predominantly petroleum imports. It is noteworthy, however, that for mineral fuels both countries have large and growing petroleum-refining industries, with substantial exports of refined petroleum, which reduces the net imports of mineral fuels. In 2006, Chinese net imports of mineral fuels were \$71.2 billion compared with \$89.0 billion of gross imports, while Indian net imports were \$43.0 billion compared with gross imports of \$61.9 billion.

The fourth and most noteworthy relationship is the sharp contrast between the very large Chinese trade surplus of \$177.5 billion in 2006, which rose to \$262.2 billion in 2007, and the moderately large Indian trade deficit of \$56.8 billion in 2006. This contrast has important consequences for the economic and policy courses ahead for both countries, as described in Chapters 5 and 9. The Chinese merchandise trade surplus fed into a current account surplus of \$250 billion in 2006, or 9.5 percent of GDP, which is projected to rise to \$400 billion or 12 percent of GDP in 2007. This huge external surplus presents a major adjustment challenge for China and for the international economic system more broadly. The \$59.3 billion Indian merchandise trade deficit, in contrast, is offset by a rapidly growing surplus in trade in services, large remittances, and a very large net inflow of long-term capital. In other words, China faces a daunting external adjustment problem, while Indian external accounts are more or less in balance.

Table 3-1
Chinese and Indian Merchandise Trade
($ billions)

	2000	**2004**	**2005**	**2006**
Exports				
China				
Total	249.2	593.4	762.0	969.1
Agriculture	26.5	22.5	26.5	27.3
Mineral fuels	7.9	14.5	17.6	17.8
Industrial raw materials	4.5	2.1	3.3	7.9
Manufactures	223.8	554.3	714.6	916.1
India*				
Total	44.6	83.5	103.1	126.4
Agriculture	9.1	8.6	13.0	15.7
Mineral fuels	0.9	7.1	11.9	18.9
Industrial raw materials	1.9	4.6	5.8	6.4
Manufactures	32.7	63.2	72.5	85.3
Imports				
China				
Total	225.1	561.4	660.1	791.6
Agriculture	48.1	21.8	22.2	14.9
Mineral fuels	20.6	48.0	64.1	89.0
Industrial raw materials	20.0	19.1	28.2	83.1
Manufactures	178.4	472.5	545.6	604.6
India*				
Total	50.5	111.5	149.2	185.7
Agriculture	2.8	4.6	5.5	7.1
Mineral fuels	17.5	34.8	50.3	61.9
Industrial raw materials	1.0	1.7	3.3	7.6
Manufactures	29.2	70.4	90.1	109.2

*Financial year, April of the listed year through March of the following year.
Sources: China's Customs Statistics; and Indian Department of Commerc Export Import Data Bank.

Table 3-2 breaks down trade in manufactures by principal sector for 2006. For China, the largest export sectors are electrical and nonelectrical machinery and equipment, together amounting to $414.0 billion, or 45 percent of the total. The next largest sectors are textiles and apparel at $138.1 billion, chemicals, including pharmaceuticals, at $37.8 billion, and precision instruments at $32.6 billion. For India, the largest export sectors are precious stones at $16.1 billion, chemicals, including pharmaceuticals, at $12.8 billion, and textiles and apparel at $11.5 billion.

Table 3-2
Chinese and Indian Trade in Manufactures
($ billions, 2006)

	China	India*
Exports		
Manufactures, Total	916.1	85.3
Chemicals, including pharmaceuticals	37.8	12.8
Textiles and apparel	138.1	11.5
Footwear	26.3	1.2
Precious stones	6.9	16.1
Iron and steel	25.1	9.0
Electrical machinery and equipment	227.4	4.1
Non-electrical machinery and equipment	186.6	5.1
Autos and parts	22.4	3.8
Aircraft	1.3	0.0
Precision instruments	32.6	0.8
Other	211.6	20.9
Imports		
Manufactures, Total	604.5	109.2
Chemicals, including pharmaceuticals	56.2	14.7
Textiles and apparel	25.7	0.8
Footwear	0.8	0.1
Precious stones	4.6	22.6
Iron and steel	20.0	8.7
Electrical machinery and equipment	219.0	14.6
Non-electrical machinery and equipment	109.2	18.6
Autos and parts	17.0	1.4
Aircraft	10.9	5.3
Precision instruments	58.8	3.1
Other	82.3	19.3

*Financial year, April 2006 through March 2007.
Sources: China's Customs Statistics; and Indian Department of Commerce, Export Import Data Bank.

As a first approximation of low versus high technology industries, the textiles and apparel, footwear, and precious stones sectors are clearly lower tech, while six of the other listed sectors—chemicals, electrical and non-electrical machinery and equipment, autos and parts, aircraft, and precision instruments—are in the higher tech category. On this basis, $500.3 billion, or 55 percent, of Chinese exports, would be high tech versus $171.3 billion, or 19 percent, low tech. For India the high-tech sectors amount to $26.6 billion, or 31 percent, compared with $28.8 billion, or 36 percent, for the three lower tech sectors. All of these figures would be somewhat higher if the unlisted "other" category were also distributed on a higher versus lower tech basis. In any event, Chinese exports are now predominantly high-tech, on a three to one ratio relative to lower tech industries, based on the listed sectors, while Indian exports are about 10 percent higher for the lower tech industries.

On the import side, the higher technology sectors are even more dominant, reflecting the fact that China and India import very little apparel and footwear, although both countries do import substantial textiles for fabrication into apparel for reexport. The six listed higher tech sectors amounted to $460.2 billion, or 76 percent, of total manufactured imports for China, and $57.7 billion, or 53 percent, for India. The Indian percentage is much lower principally because of the large $22.6 billion of imports of precious stones, largely worked into jewelry for reexport.

Table 3-3 presents Chinese and Indian trade with the United States, the EU, Japan, and each other for 2004-2006. For Chinese exports in 2006, the United States and the EU were numbers one and two, at $203.5 billion and $182.0 billion, with Japan a distant third at $91.6 billion. During 2007, not shown in the table, Chinese exports to the EU rose faster than and surpassed those to the United States. Imports, in contrast, are skewed in the opposite direction: Japan $115.7 billion, the EU $90.3 billion, and the United States $59.2 billion. These contrasts result in an even more striking disparity in the trade balances: a $144.3 billion surplus with the United States, a $91.7 billion surplus with the EU, and a $24.1 billion deficit with Japan. These diverse trade imbalances are substantially understated, moreover, by the very large Chinese trade surplus with Hong Kong, not shown in the table, at $144.6 billion in 2006, over 90 percent of which is reexported, primarily to the United States, the EU, and Japan. An adjustment is made for these Hong Kong transshipments in Table 3-4, based on the shares of Hong Kong reexports to the three destinations. The result is an increase in the Chinese surpluses to $180.7 billion for the United States and $125.5 billion for the EU, and a decline in the deficit with Japan to $11.1 billion.[48]

Indian exports are led by the EU market, at $26.8 billion in 2006, followed by the United States at $18.9 billion. Exports to China are much smaller, at $8.3 billion, but still higher than the $2.9 billion of exports to Japan. Indian imports are also heavily skewed in favor of the EU: the EU $29.8 billion, China $17.5 billion, the United States $11.7 billion, and Japan $4.6 billion. A larger share of Indian imports

---

[48] Even with the Hong Kong adjustment, the official Chinese figure of a $180.8 billion surplus with the United States is much smaller than the $232.6 billion U.S. figure for the bilateral deficit with China, as discussed below.

is also from other sources, primarily the large share of total imports in petroleum and industrial raw materials.

Table 3-3
Chinese and Indian Merchandise Trade by Principal Trading Partner
($ billions)

	China			India*		
	2004	**2005**	**2006**	**2004**	**2005**	**2006**
Total						
Exports	593.4	762.0	969.1	83.5	103.1	126.4
Imports	561.4	660.1	791.6	111.5	149.2	185.7
Balance	+32.0	+101.9	+177.5	-28.0	-46.1	-59.3
United States						
Exports	124.9	162.9	203.5	13.8	17.4	18.9
Imports	44.7	48.7	59.2	7.0	9.5	11.7
Balance	+80.2	+114.2	+144.3	+6.8	+7.9	+7.2
EU						
Exports	107.2	143.7	182.0	18.1	23.2	26.8
Imports	70.1	73.6	90.3	19.1	25.9	29.8
Balance	+37.1	+70.1	+91.7	-1.0	-2.7	-3.0
Japan						
Exports	73.5	84.0	91.6	2.1	2.5	2.9
Imports	94.4	100.5	115.7	3.2	4.1	4.6
Balance	-20.9	-16.5	-24.1	-1.1	-1.6	-1.7
China						
Exports	X	X	X	5.6	6.8	8.3
Imports	X	X	X	7.1	10.9	17.5
Balance	X	X	X	-1.5	-4.1	-9.2
India						
Exports	5.9	8.9	14.6	X	X	X
Imports	7.7	9.8	10.3	X	X	X
Balance	-1.8	-0.9	+4.3	X	X	X

*Financial year, April of the listed year through March of the following year.
Sources: China's Customs Statistics; Indian Department of Commerce, Export Import Data Bank.

Table 3-4
Merchandise Trade Balance Adjusted for Hong Kong Transshipments
($ billions, 2006)

	Official Chinese Figures	**Adjusted for Hong Kong Transshipments**
United States	+144.3	+180.7
EU	+91.7	+125.5
Japan	-24.1	-11.1

Sources: China's Customs Statistics; and Hong Kong Census and Statistics Department.

A principal observation from Table 3-3 is the far stronger position of the EU compared with the United States with respect to both the Chinese and Indian markets. In 2006, China had $90.3 billion of imports from the EU compared with $59.2 billion from the United States, while India had $29.8 billion from the EU compared with $11.7 billion from the United States. These relative import levels, predominantly manufactures, are a negative indicator for U.S. export competitiveness vis-à-vis Europe in the rapidly growing Chinese and Indian markets. The rise of the euro rate to the dollar, however, should have a favorable impact on the U.S. market shares.

Another disturbing observation for U.S. exports to India is the rapid rise of Indian imports from China, rising from rough parity with the United States in 2004 to a much higher $17.5 billion in 2006 compared to only $11.7 billion from the United States.

A final related observation is the very rapid rise in Chinese-Indian trade, strongly favoring Chinese exports to India. Chinese exports to India rose from $5.9 billion in 2004 to $14.6 billion in 2006, or by 147 percent, while Chinese imports from India were up from $7.7 billion to $10.3 billion, or by 34 percent. In 2006 alone, Chinese exports to India were up by 60 percent, compared with only a 5 percent increase in imports from India. As a result, the Chinese bilateral trade balance, based on the Chinese statistics, shifted from a $1.8 billion deficit in 2004 to a $4.3 billion surplus in 2006; based on the Indian statistics, the deficit with China rose from $1.5 billion to $9.2 billion.

Table 3-5 elaborates the export competitive relationships by presenting Chinese and Indian imports of the six high technology sectors listed in Table 3-1, for 2005, by source from the United States, the EU, Japan, and each other. The six sectors make up the large majority of imports from the sources, as shown in the bottom two lines of the table, ranging from 62.4 to 73.8 percent, except for Chinese imports from India, at only 14.3 percent. In effect, this table presents a picture of export competitiveness for the principal high technology industries in the rapidly growing Chinese and Indian markets, among the five trading partners.

The most striking observation from Table 3-5 is again the larger market shares of the EU compared with the United States, in both China and India, in all of the sectors except aircraft, which is a standoff between Boeing and Airbus. The EU lead is largest for electrical machinery and equipment, nonelectrical machinery and equipment, and autos and parts, and smaller for chemicals and precision instruments. For all six sectors, Chinese imports were $53.6 billion from the EU compared with $30.4 billion from the United States, and corresponding Indian imports were $16.7 billion and $8.1 billion.

Table 3-5
Chinese and Indian* Imports of Advanced Technology Industries
($ billions, 2006)

Imports from	United States	EU	Japan	China	India
Chemicals, including pharmaceuticals					
China	5.5	6.7	8.9	X	1.0
India	2.1	2.4	0.5	2.5	X
Electrical machinery and equipment					
China	8.5	12.7	29.8	X	0.1
India	1.3	3.5	0.5	4.2	X
Non-electrical machinery and equipment					
China	8.3	22.9	21.6	X	0.2
India	1.9	6.6	1.5	3.2	X
Autos and parts					
China	0.9	4.1	4.3	X	0.0
India	0.0	0.5	0.2	0.2	X
Aircraft					
China	3.4	3.0	0.0	X	0.0
India	2.0	2.6	0.0	0.0	X
Precision instruments					
China	3.8	4.2	8.7	X	0.1
India	0.8	1.1	0.3	0.2	X
Total, listed industries					
China	30.4	53.6	73.3	X	1.4
India	8.1	16.7	3.0	10.3	X
Listed industries as a percent of total Imports					
China	62.4	72.8	72.9	X	14.3
India	69.2	56.0	65.2	58.9	X

*Financial year, April through March of the following year.
Sources: China's Customs Statistics; and Indian Department of Commerce, Export Import Data Bank.

The Japanese market shares are mixed, generally higher than the EU and the United States in China, except for aircraft, while much lower in India. For all six sectors, Chinese imports from Japan were $73.3 billion, compared with $53.6 billion from the EU and $30.4 billion from the United States. Corresponding Indian imports were EU $16.7 billion, the United States $8.1 billion, and Japan $3.0 billion. A final observation concerns Indian imports from China, which for all six high-tech sectors, in 2005, were $10.3 billion, and substantially higher than the $8.1 billion of imports from the United States.

Table 3-6 presents Chinese and Indian trade in services, with a greatly different picture than for merchandise trade, in favor of India. Total services trade, shown in the top half of the table, consists of three principal categories: transportation (mainly ocean shipping and airlines); travel (mainly tourism and business travel); and other commercial services (including IT, engineering, telecommunications, and financial). Other commercial services, in the bottom half of the table, have been growing the fastest, particularly for Indian exports. The Indian software and other business services sector is discussed later in the chapter, but the basic figures for overall other business services show the basic picture. Indian exports tripled from 2003 to 2006, from $19.5 billion to $58.3 billion, moving far ahead of China in the process, with exports of $21.1 billion and $36.5 billion, respectively. For total trade in services, in 2006, India had a $16.0 billion surplus compared with an $8.9 billion deficit for China, while for other commercial services the $26.9 billion India surplus compared with a $5.1 billion deficit for China.

Table 3-6
Chinese and Indian Trade in Services
($ billions)

	2003	**2004**	**2005**	**2006**
Total Services				
Exports				
China	46.4	62.1	73.9	91.4
India	25.0	37.2	54.4	73.8
Imports				
China	54.9	71.6	83.2	100.3
India	21.6	36.5	49.2	63.7
Trade Balance				
China	-8.5	-9.5	-9.3	-8.9
India	+3.4	+0.7	+5.2	+16.0
Other Commercial Services				
Exports				
China	21.1	24.2	29.2	36.5
India	19.5*	30.9	43.8	58.3
Imports				
China	21.4	27.9	33.0	41.6
India	14.7*	22.1	26.6	31.4
Trade Balance				
China	-0.3	-3.7	-3.8	-5.1
India	+4.8	+8.8	+17.2	+26.9

*Estimated, based on the share of total in 2004.
Source: WTO, *International Trade Statistics*, 2004-2007.

The far greater relative importance of services exports for India is encapsulated in Table 3-7, which compares exports of manufactures with exports of other commercial services for the two countries in 2006. Chinese exports of manufactures were more than ten times larger than Indian exports, while for other commercial services India is well ahead. Other commercial services as a percent of manufactures, on the bottom line of the table, shows 68 percent for India compared with only 4 percent for China, and this 68 percent figure will rise further to the extent Indian growth in other commercial services exports continues to outstrip growth in manufactured exports. In effect, exports of other commercial services have been the driving force for Indian advanced technology export growth, just as the manufacturing sector was for China beginning in the mid-1990s.

Unfortunately, official statistics on trade in services are not broken down by country of import source or by export destination, although some indications of the U.S. share are noted in the final section of this chapter.

Table 3-7
Exports of Manufactures and Other Commercial Services, 2006
($ billions)

	China	India
Manufactures	916.1	85.3
Other Commercial Services	36.5	58.3
Other Commercial Services as Percent of Manufactures	4.0	68.0

Sources: Tables 3-1 and 3-6.

Chinese Manufactured Exports Go High Tech

The structural shift in Chinese exports from low to high technology industries was shown in broad terms in Table 2-3, and is elaborated and focused more sharply in Table 3-8. Sections A and B of the table are selected categories of higher tech industries and products, which China began to include in its trade statistics in 2002 to emphasize the central policy objective of promoting export-oriented advanced technology industry. These two categories of trade statistics require a background explanation about the distinction in trade statistics between "high-tech industries" and "high" or "advanced technology" products. The high-tech industries category includes all products within the designated industries, and the most commonly used listing, over several decades, has been that of the OECD. The OECD, in fact, classifies industries as "high," "medium high," and "low" technology. The high or advanced technology products category, in contrast, is limited to those specific products within that have the highest R&D and engineering content. The United States adopted its advanced technology product, or ATP, classification in the late 1980s, of which more below, and China did likewise, in 2002, with its "high-tech products" category. The U.S. and Chinese advanced and high technology product categories are very similar, and there has been considerable discussion of them between experts from the two countries.[49]

In Table 3-8, "mechanical and electrical products" is an industry classification, and includes most, although not all, industries designated as high technology by the OECD, while "Hi-tech products" is the more narrowly drawn selection of products, principally but not all from the mechanical and electrical products category. The third section lists the principal labor-intensive export sectors: textiles and apparel, footwear, toys and sporting goods, and furniture.

The most striking observation from Table 3-8 is the predominance of the high technology over the labor-intensive exports. Exports of mechanical and electrical products were $549.4 billion in 2006, or 60 percent of total manufactured exports, and almost three times larger than the $203.4 billion of labor-intensive products. Even the more selective high-tech products category recorded exports of $281.4 billion, well above the level for labor-intensive products.

The strong trend toward higher tech exports is also clear from the respective growth rates. In 2006, exports grew by 29 percent both for mechanical and electrical products and for high-tech products, compared with 20 percent growth for the labor-intensive products. The trade surpluses were also up far more sharply, by 466 percent and 734 percent, from 2004 to 2006, for the two higher tech categories, compared with a 47 percent increase for the labor-intensive category. Chinese apparel and textile exports receive widespread media attention for gaining global market share, but their share of Chinese manufactured exports are in secular decline, down from 17 percent in 2004 to 16 percent in 2005 to 15 percent in 2006.

One critical dimension of Chinese high-tech exports that does not appear in the table is the preponderant share produced by foreign invested firms. For total Chinese exports, the share by foreign firms, wholly owned or joint ventures, rose from 13 percent in 1990 to 32 percent in 1995 to 58 percent in 2007. The foreign firm share for low technology industries, however, is very low. Exports of apparel footwear, toys, and furniture are mostly by Chinese firms. High,

[49] See Michael Ferrantino, *et al.*, "Classification and Statistical Reconciliation of Trade in Advanced Technology Products: The Case of China and the United States" (United States International Trade Commission, September 2007).

Table 3-8
Chinese Trade in Machinery and Electrical, Hi-Tech, and Labor-Intensive Products
($ billions)

	Exports			Imports			Trade Balance		
	2004	2005	2006	2004	2005	2006	2004	2005	2006
Mechanical and Electrical Products, Total	323.4	426.7	549.4	301.9	350.4	427.7	+21.5	+76.3	+121.7
Metal products	22.4	29.5	38.8	6.1	7.0	8.7	+16.3	+22.5	+30.1
Machinery	118.1	149.7	186.6	91.6	96.4	109.2	+26.5	+53.3	+77.4
Electric and electronic products	129.7	172.3	227.4	142.1	174.8	219.0	-12.4	-2.5	+8.4
Transport equipment	21.0	28.4	38.4	19.5	19.9	29.7	+1.5	+8.5	+8.7
Instrument and apparatus	16.2	25.5	32.6	40.2	50.0	58.8	-24.0	-24.5	-26.2
Other	15.9	21.3	25.6	2.5	2.3	2.3	+13.4	+19.0	+23.2
Hi-Tech Products, Total	165.5	218.2	281.5	161.4	197.7	247.3	+4.1	+20.5	+34.2
Biotechnology	0.2	0.3	0.3	0.1	0.1	0.2	+0.1	+0.2	+0.1
Life science	3.2	4.6	6.3	3.8	4.6	5.1	-0.6	0.0	+1.2
Opto-electronics	3.8	7.2	7.1	3.2	3.4	4.0	+0.6	+3.8	+3.1
Computer and telecommunications	136.4	177.1	224.9	50.7	60.3	70.7	+85.7	+116.8	+154.2
Electronics	18.4	24.5	36.0	77.2	100.9	130.2	-58.8	-76.4	-94.2
Computer integrated manufacturing	1.5	2.1	2.9	17.4	16.6	19.7	-15.9	-14.5	-16.8
Materials	0.7	0.9	1.3	2.3	2.8	4.0	-1.6	-1.9	-2.7
Aerospace	1.0	1.4	2.4	6.4	8.7	13.2	-5.4	-7.3	-10.8
Other technology	0.3	0.3	0.3	0.3	0.2	0.2	0.0	+0.1	+0.1
Labor-Intensive Products, Total	138.2	170.1	203.4	18.5	18.8	27.9	+119.7	+151.3	+175.5
Textiles and apparel	95.3	115.3	138.1	16.8	17.1	25.7	+78.5	+98.2	+112.4
Footwear	15.2	19.1	21.8	0.5	0.5	0.6	+14.7	+18.6	+21.2
Toy and sporting goods	15.1	19.1	22.6	0.5	0.6	0.8	+14.6	+18.5	+21.8
Furniture	12.6	16.6	20.9	0.7	0.6	0.8	+11.9	+16.0	+20.1

Source: China's Customs Statistics, Monthly Exports and Imports, Series No. 196, Tables 3, 4, 13, and 14.

technology exports, in contrast, are almost all by foreign firms. Based on the OECD high technology industry classification, 88 percent of Chinese high technology exports are by foreign firms,[50] while for exports from within the special economic zones, where much of the

[50] *OECD Reviews of Innovation Policy: China*, Synthesis Report (OECD, September 2007), p. 15.

electronics and information technology products are processed for export, the foreign firm share is 95 to 100 percent.[51] This rapid growth of high technology exports by foreign firms, however, could be put at risk by government actions to favor Chinese firms in the context of the independent innovation strategy and to revalue the yuan. Both actions could lead to a shift of new export-oriented investment to other locations, including in India.

The Fading Export Platform Issue

One controversial subject concerning Chinese exports of advanced technology products, introduced earlier, has been the export platform issue, whereby imported components comprise a substantial share of the value of the exported final product. This issue has received prominent attention in recent years for the electronics, information technology, and telecommunications sectors, involving imported components from Taiwan, South Korea, and Japan, in particular. The issue is of special importance in view of the preponderant role of such foreign firm production in Chinese high technology exports described in the previous paragraph.

The share of such imported components, however, is not nearly as large as sometimes reported, and the share of Chinese value added is rising steadily. Foreign firms are increasing their sourcing of components from domestic suppliers, which is reflected in the rapid growth in the trade surplus in the high technology sectors. Japanese automobile companies are increasing their Chinese value added to as high as 90 percent, largely because Japanese parts producers have been shifting production to China. For the information technology and electronics sectors, as described earlier, Taiwanese firms are shifting increasing shares of value added to China, including for engineering and product design.

Unfortunately, there are no hard figures on the average Chinese value added in exports of high technology industries. Over half of these exports now come from export processing zones, which provide special incentives, and Chinese value added within the zones has been estimated at 30 to 35 percent, but this does not take account of the rising share of components from outside the zones from domestic suppliers, or of other Chinese value added, such as capital costs.

The most direct indicator of the rising share of Chinese value added in exports is from the trade statistics themselves. From 2002 to 2007, Chinese imports doubled, but exports tripled and the trade surplus

[51] Zhi Wang and Shang-Jin Wei, "The Rising Sophistication of China's Exports: Assessing the Roles of Processing Trade, Foreign Invested Firms, Human Capital, and Government Policies" (paper presented at the Carnegie Foundation, Washington, DC, September 26, 2007), p. 5.

increased by ten times.[52] For the electronics, information technology, and telecommunications sectors, which is where the export platform issue discussion has centered, the trade figures for 2003-2007 are presented in Table 3-9. The Chinese trade surplus in these sectors increased by almost 500 percent, from $26.5 billion to $148.7 billion, while the surplus as a percentage of imports rose from 18 to 34 percent. This means that if all imports in the sector were reassembled into exports, the Chinese value added would have been 34 percent. But, of course, many and perhaps most imports in the sector are for the domestic Chinese market. Of the $296.9 billion of imports in 2007, 10 to 20 percent were in final products for the domestic market. Another 47 percent were semiconductors and electronic integrated circuits, a large share of which is destined for wide-ranging products on the domestic market. The remaining 30 to 40 percent of imports of components were probably mostly, but not all, for export assembly.

Table 3-9
Chinese Trade In the Information Technology and Telecommunications Sector
($ billions)

	2003	2004	2005	2006	2007*
Exports	150.0	215.6	281.1	359.8	441.6
Imports	123.5	164.7	202.9	251.0	292.9
Trade balance	+26.5	+50.9	+78.2	+108.8	+148.7
Trade surplus as a percent of exports	18%	24%	28%	30%	34%

*Projected, based on January-October.
Source: Table 3-2; the three listed categories of office machines and equipment, telecommunications equipment, and electrical machinery and parts, or SITC 75-77, for 2004-2006.

If half of the sectoral imports were destined for the domestic market, average Chinese value added for exports in 2007 would have been 66 percent (i.e., $148.7 billion of imports for $441.6 billion of exports, or 34 percent import content). Moreover, the domestic share will continue to rise as export industry turns further to domestic suppliers of components and the trade surplus continues to grow. The semiconductor sector, in particular, is likely to show a large shift to domestic suppliers as new Chinese production comes on stream, and

[52] One study, based on input-output analysis, estimated a 37 percent average Chinese value added for total Chinese exports, but this estimate was based on trade in 2002, and the percentage would be much higher if based on 2007 trade because the input-output coefficients have changed as a result of the soaring trade surplus. See Lawrence Lao, "Estimates of U.S.-China Bilateral Trade Balances in Terms of National Value-Added" (presentation at the Australian National University, Canberra, December 11, 2006).

the sector could even reverse into trade surplus over the next several years. Chinese exports of semiconductors and electronic integrated circuits increased from $3.9 billion in 2000 to $8.1 billion in 2003 and $29.0 billion in 2007.

There will continue to be substantial imported content in Chinese high technology exports, but the share of Chinese value added is probably now 60 percent or higher, and continues to rise. This is similar to other export industries around the world, increasingly engaged in cross-border supply chain management. It is in the same range, for example, as Boeing aircraft, the largest U.S. export product to China. In conclusion, the export platform issue, in recent years, has been greatly overstated. Chinese value added in electronics, information technology, and telecommunications exports is not very much below average for global high technology industry trade and is steadily rising.

Indian Exports of Software and Other Business Services

Surging Indian exports of "other commercial services" were shown in Tables 3-6 and 3-7, rising to $58 billion in 2006, or 68 percent of exports of manufactures. If recent export growth rates continue, such services exports would surpass manufactures in about five years. Observers acknowledge this extraordinary export performance, but they often temper their praise by commenting that the success is limited to a few enclaves, such as Bangalore and Hyderabad, and is thus having relatively small impact on the overall course of the predominantly poor Indian economy. This downplay misses the mark, however, and is misleading. The rapid growth of this sector is having a much broader impact on the industrial modernization of India. The sector is, in fact, a growing force for far-reaching technological change throughout the economy and within Indian society as a whole, with surging exports the catalytic leading edge.

The creation and development of the software and other business services sector, or the IT services sector for short, is an extraordinary story of bold, risk-taking entrepreneurs.[53] There is a striking parallel

[53] This sector has various labels. The Indian industry association, NASSCOM, calls it the information technology and business practices outsourcing sector, or IT-BPO, although this name implies that all services are exported, or outsourced, whereas in fact a significant and growing share of these services are for the Indian market. Software is explicitly included in the definition used in this study because the large majority of Indian software is in software services, the application of existing patented programs, rather than in newly patented software, which is classified in trade data as "goods" rather than "services." The WTO trade classification, "other commercial services," is somewhat broader than the IT services sector discussed here.

between the early entrepreneurial Silicon Valley leaders—Hewlett and Packard at the outset, followed by Steve Jobs and Bill Gates—and the big three of IT services companies in India: Tata Consultancy Services (TCS), Infosys, and Wipro. TCS is part of the Tata Group, with a history of entrepreneurial boldness dating back to the late 19th century, and the family scion, Ratan Tata, was far ahead in creating TCS in 1965, when IT services were in their unpromising infancy, and in building it into a software and engineering giant with $5 billion of revenue in 2007. Infosys was founded in 1981 by an engineer, Narayana Murthy, currently chairman, with $250 of start-up capital and a bold vision, and he has likewise created a $3 billion business. Wipro was founded in 1985 by current CEO Azim Premji with modest start-up capital, and has also risen to a $3 billion firm today. All three have annual growth in revenue and profits in the range of 30 to 40 percent.

The overall IT services sector has grown dramatically and is broadening in the scope of services provided, including within the Indian economy, although exports still dominate, accounting for 65 to 70 percent of revenue. A growing number of modern industrial parks house clusters of Indian and multinational company "campuses," many engaged in engineer-intensive R&D. It is hard to get a precise account of the size and international performance of the sector in view of its rapid growth and the time lag and questionable coverage of official statistics. The most current source of statistics is the business organization the National Association of Software and Service Companies, or NASSCOM, and its estimates and projections confirm the basic picture of rapid, export-led growth.[54] Data from this source indicate that the Indian IT services sector grew tenfold, from $4.8 billion in 1997 to $47.8 billion in 2006,[55] with a corresponding rise in the share of Indian GDP from 1.2 to 5.4 percent. Direct employment in 2006 was 1.6 million, with indirect employment at least as large. By December 2006, 440 Indian companies had acquired quality certification, of which 90 were certified at SEI CMM Level 5, a higher number than in any other country. Recent annual revenue growth for the sector has been 28 percent, led by exports at over 30 percent, and these rates of growth are projected to continue over the next several years. Exports are projected to rise from $31 billion in 2006, of which

---

[54] See *IT Strategic View 2007* (NASSCOM, January 2007), at www.nasscom.org.

[55] As elsewhere, these figures are on a financial year basis, from April of the cited year to March of the following year, thus with a nine month overlap with the calendar year. The NASCOMM source, like official statistics, cites the year at the end of the financial year in March, with only a three-month overlap with the calendar year. The WTO figures in Table 3-6, in contrast, are on a calendar year basis and are therefore not comparable.

$18 billion is in IT and $13 billion is in business services, to $60 billion in 2009.

What this all means is that the rapid growth of the market- and free trade-oriented IT services sector in India presents a greatly different model for industrial modernization and growth than was the case in the United States and China, where manufacturing industry, including IT hardware, was the principal driving force. The Indian IT services sector is not only leading the way for export-driven growth, but, to an increasing extent, is the catalyst for modernization and growth throughout the domestic economy. The recent rapid increase in growth in the manufacturing sector and the emerging surge in infrastructure project investment all draw heavily on advanced technology software development and engineering services provided by the IT services sector, including the participation by multinational companies. The critical need for higher quality education, with outreach to the provincial poor, will also require creative software and other business organizing services at all levels. The strengthening of the democratic process and the rule of law in India, discussed in Chapter 5, is yet another beneficiary of the application of new software to provide more prompt and transparent information within the legal system. In effect, the heavy and slow-moving hand of the license Raj is being challenged, and progressively replaced, by software and other products emanating from the IT services sector.

International attention has highlighted the trade performance of the Indian IT services sector, but even this is not adequately observed because of the time lag and lack of detail for official data on services as compared with merchandise trade. Most media reports on trade are limited to merchandise trade, and seldom note that services trade is excluded, which makes a big difference for India. Data for trade in services from the WTO, the only readily available official source, are one to two years behind the time, a lag that is a serious shortcoming for Indian exports of services growing at 30 percent or more per year. And unlike merchandise trade, services trade data are not broken down by country source of imports and destination of exports. Official figures do not disclose, for example, that the large majority of Indian IT services exports, in the order of 60 to 80 percent, has been to the United States, the most open market with wide-ranging business connections through the Indian diaspora, although this share is gradually declining as the European, Asian, and other markets are more actively developed.

U.S. Bilateral Trade: High Growth and Deepening Deficits

U.S. bilateral trade with China and India has two dominant characteristics, one very positive for the United States and the other deeply disturbing. The positive characteristic is that both countries are high growth markets for U.S. exports, with enormous longer term potential, and with large mutual economic gains from imports as well as exports. The disturbing characteristic is that U.S. trade with both countries is in very large and growing deficit, in the case of China an unprecedented trade imbalance between the two largest trading nations.[56] These export opportunities and trade deficits are having substantial impact on the central global economic challenges facing the United States of restoring export competitiveness and maintaining leadership in technological innovation. The basic bilateral trade figures are presented here as the empirical setting for the U.S. policy response to these challenges as recommended in Chapters 8 through 10. Merchandise trade is addressed first, in greater detail because the data are more readily available, and then trade in services, the data for which are woefully inadequate.

Bilateral merchandise trade with China and India is summarized in Table 3-10, from 2004 to 2006, broken down by broad sector. The far larger trade with China stands out: U.S. exports to China in 2006 were more than five times larger than to India, at $55.2 billion compared with $10.1 billion, and imports were 13 times larger, at $287.8 billion compared with $21.8 billion. The predominance of trade in manufactures, described earlier in greater detail, is also evident: in 2006, for China, 72 percent of U.S. exports and 98 percent of imports were manufactures; for India, the corresponding figures were 87 percent of exports and 92 percent of imports.

U.S. export growth, a positive characteristic of the trade relationships, is high for both countries. U.S. exports to China were up 59 percent from 2004 to 2006, to $55.2 billion. For India, exports were up by 66 percent to $10.1 billion. Growth in manufactured exports was comparable, up by 59 percent to $39.7 billion for China, and by 69 percent to $8.8 billion for India.

The disturbing characteristic of the growing trade deficits, however, is even more striking. In 2006, U.S. merchandise imports from China, at $287.8 billion, were more than five times larger than exports, with a resulting trade deficit of $232.6 billion. For India, imports, at $21.8 billion, were more than twice as large as exports, with a resulting deficit of $11.3 billion. The $232.6 billion deficit with China is much larger than the $144.3 billion deficit in the Chinese statistics shown in Table 3-3. Part of the difference comes from the

---

[56] Germany is often cited as number one or two, but over 60 percent of German trade is within the EU and is thus not part of the WTO trading system. EU-extra trade is the more comparable measure.

Table 3-10
U.S. Merchandise Trade With China and India by Sector*
($ billions)

	2004	**2005**	**2006**	**Percent Change 2004-2006**
China				
Exports, total	34.7	41.8	55.2	+59
Agriculture	1.4	1.2	1.6	+14
Industrial raw materials	8.1	9.9	13.7	
Mineral fuels	0.2	0.1	0.2	0
Manufactures	25.0	30.6	39.7	
Imports, total	196.7	243.5	287.8	+46
Agriculture	2.4	2.8	3.7	+54
Industrial raw materials	1.0	1.3	1.5	+50
Mineral fuels	1.0	1.0	1.2	+20
Manufactures	192.3	238.4	281.4	+46
Trade balance, total	-162.0	-201.7	-232.6	
Agriculture	-1.0	-1.6	-2.1	
Industrial raw materials	+7.1	+8.6	+12.5	
Mineral fuels	-0.8	-0.9	-1.0	
Manufactures	-167.3	-207.8	-241.7	
India				
Exports, total	6.1	8.0	10.1	+66
Agriculture	0.2	0.2	0.3	+50
Industrial raw materials	0.4	0.6	0.6	+33
Mineral fuels	0.3	0.4	0.4	+25
Manufactures	5.2	6.6	8.8	+69
Imports, total	15.6	18.8	21.8	+40
Agriculture	1.0	1.0	0.9	-11
Industrial raw materials	0.2	0.3	0.4	+50
Mineral fuels	0.3	0.7	0.4	+25
Manufactures	14.1	16.8	20.1	+43
Trade balance, total	-9.5	-10.8	-11.7	
Agriculture	-0.8	-0.8	-0.6	
Industrial raw materials	+0.2	+0.3	+0.2	
Mineral fuels	0.0	-0.1	0.0	
Manufactures	-8.9	-10.2	-11.3	

*The sectors are by Standard International Trade Category (SITC): agriculture 0, 1, 4; industrial raw materials 2; mineral fuels 3; and manufactures 5-9. SITC is "classified elsewhere," which would mean, on average, over 90 percent in manufactures and, in any event, this category accounts for less than 2 percent of total trade.
Source: U.S. Bureau of the Census.

Hong Kong adjustment, in Table 3-4, which brings the Chinese figure up to $180.7 billion. The remaining difference comes largely from Chinese exports to other third countries and free ports for reexport to the United States, with or without further processing, which the United States considers to be imports from China.

For China, the trade deficits are even larger for manufactures than for total trade. In 2006, the $241.7 billion deficit for manufactures was $9.1 billion larger than the $232.6 billion total deficit. This principally reflects the $13.7 billion U.S. surplus in industrial raw materials, such as ferrous and nonferrous basic metals, waste, and scrap, pulp and waste paper, and cotton textile fiber, all of which are inputs for Chinese manufacturing industry.

Another significant observation from Table 3-10, in view of the great discord over agricultural trade in the WTO Doha Round negotiations, is that only 3 percent of U.S. exports to both China and India were in agriculture, and there are small deficits with both countries, the deficit with China rising from $1.0 billion in 2004 to $2.1 billion in 2006.

The linkage between the massive U.S. trade deficit with China and a decline in U.S. technological leadership is more sharply focused in the advanced technology products, or ATP, category of U.S. trade statistics. This special category was developed in the mid-1980s to highlight industrial products with the highest share of R&D and engineering content, at a time when concern was raised over the Japanese challenge to U.S. technological leadership.[57] Approximately 600 products are contained in the ATP listing, accounting for about one-quarter of U.S. trade in manufactures, and the product selection is periodically revised by industry experts at the Bureau of the Census. The initial years of ATP trade account were reassuring, with a U.S. global trade surplus ranging mostly from $20 billion to $35 billion per year. Then in 2000 the surplus declined abruptly to $5.3 billion, beginning a steady downward trend into a deficit of $16.6 billion in 2002 and $52.0 billion in 2006.

This sharp decline into deficit for ATP trade is principally accounted for by the rapidly rising bilateral deficit with China. In 1998, ATP trade with China was in balance, with $6.1 billion of trade in each direction. The balance then shifted into a growing U.S. deficit, as shown in Table 3-11: $11.8 billion in 2002, $36.3 billion in 2004,

[57] The ATP origins, as well as an assessment of U.S. ATP trade by sector and region, are contained in Ernest H. Preeg, *The Threatened U.S. Competitive Lead in Advanced Technology Products (ATP)* (Manufacturers Alliance/MAPI, March 2004).

and $55.1 billion in 2006. In 2006, the bilateral ATP deficit with China was thus larger than the global deficit of $52.0 billion.

Table 3-11
U.S. ATP Trade With China and India
($ billions)

	2002	2003	2004	2005	2006
China					
U.S. exports	8.3	8.3	9.4	12.3	17.6
U.S. imports	20.1	29.3	45.7	59.2	72.7
Trade balance	-11.8	-21.0	-36.3	-46.2	-55.1
India					
U.S. exports	1.3	1.3	1.5	2.1	3.3
U.S. imports	0.3	0.3	0.3	0.5	0.7
Trade balance	+1.0	+1.0	+1.2	+1.6	+2.6

Source: U.S. Bureau of the Census.

ATP trade with India, also shown in Table 3-11, had a small but growing U.S. trade surplus of $2.6 billion in 2006. This surplus should continue to grow, probably at a rapid rate, in parallel with the high growth of industrial production and infrastructure investment in India, which involves a large amount of imported advanced technology products not produced domestically.

ATP trade with China in 2006 is broken down by the 10 sectors provided by the Bureau of the Census in Table 3-12. Almost 90 percent of imports are in the information and communications sector, at $64.4 billion, which is in keeping with the Chinese top priority for development of this sector. The other two sectors with large imports are the closely related opto-electronics and electronics sectors, at $4.4 billion and $2.3 billion, respectively. The largest sectors of U.S. exports are aerospace, principally commercial jet aircraft, and electronics, each with $6.0 billion, followed by information and communications and flexible manufacturing, at $3.2 billion and $1.1 billion, respectively. In view of the dominant position of the information and communications sector, a further statistical breakdown by subsector would be useful, as was done on one occasion in the 1990s by the National Science Foundation, and which could be provided in the Bureau of the Census monthly trade report with a simple adjustment in its software program.

U.S. bilateral trade in services is also important and growing rapidly, particularly for Indian exports of other commercial services. Unfortunately, official statistics are woefully inadequate for an accurate, timely assessment. The annual WTO data for trade in services, as shown earlier for China and India in Table 3-6, do not include a breakdown by trading partner. Likewise, U.S. monthly data for services trade provide no information on trade by industry and no breakdown by trading partner. Once a year, in the October issue of the *Survey of Current Business*, some country-specific data are provided for the previous year, and are therefore almost a year out of date. Such services trade data, moreover, fall into two categories in U.S. statistics: cross-border sales and purchases, such as for air and sea transportation and telecommunications, and sales and purchases through foreign affiliates, which would include the large majority of the IT and other business service exports of India to the United States. Globally, the trade through foreign affiliates is about 50 percent larger than cross-border services trade. The October 2007 *Survey,* however, provided 2006 data for cross-border services trade, with a sectoral breakout for "other private services," but only 2005 data, or almost two years behind, for trade through foreign affiliates, with no sectoral breakouts.

Table 3-12
U.S. ATP Trade With China, by Sector
($ millions, 2006)

	Exports	Imports
Total	17,627	72,709
Biotechnology	33	47
Life science	934	614
Opto-electronics	266	4,397
Information & communications	3,156	64,396
Electronics	6,012	2,338
Flexible manufacturing	1,076	397
Advanced materials	118	118
Aerospace	6,014	256
Weapons	1	99
Nuclear technology	16	48

Source: U.S. Bureau of the Census.

These limited data for U.S.-Indian bilateral trade in services are presented in Table 3-13 for 2005 and 2006. Cross border trade was close to balance, with $5.2 billion of exports and $5.0 billion of imports in 2005. A relatively small part of this was in "other private services," a category comparable to "other commercial services" in the WTO data, at $2.1 billion of exports and $0.9 billion of imports. The figures for trade through affiliates, however, are wildly out of line with Indian export figures presented earlier. The U.S. figures show only $2.4 billion of imports through affiliates, which, when added to the $0.9

billion of cross-border other private services, totals $3.3 billion. Indian export figures of other commercial services, in contrast, were $43.8 billion globally in 2005, of which, based on NASSCOM reports, about two-thirds, or $29.3 billion, went to the United States. Thus there is a ten to one discrepancy, and if the U.S. figures are believed, the whole issue of outsourcing of software, engineering and other business services to India becomes very small and inconsequential. In fact, however, the Indian figures are much closer to the truth, whereas the U.S. figures are grossly understated and highly misleading.

Table 3-13
U.S. Bilateral Trade in Services With India
($ billions)

	2005	2006
Exports		
Cross border	5.2	6.7
Other private services	2.1	3.1
Through affiliates	2.8	-
Imports		
Cross border	5.0	6.6
Other private services	0.9	1.1
Through affiliates	2.4	-

Source: *Survey of Current Business,* October 2007.

In view of the rapidly growing U.S. trade in business services with India, in particular, these highly conflicting data should be reconciled. The U.S. Government Accountability Office (GAO) provided a report on this subject in 2005, based on U.S.-Indian trade in 2002 and 2003.[58] The report showed a more than 20:1 difference between U.S. import and Indian export statistics for the business services sector. It also identified five factors that contributed to the difference: U.S. data count the earnings of temporary Indian workers for one year while Indian data include the full stay; U.S. data do not include certain software and financial services while Indian data do; U.S. data do not count Indian exports to U.S. firms located outside the United States while Indian data do; U.S. data do not count trade between foreign companies and their U.S. affiliates, due to the quality of the responses, while Indian data do; and other collection or methodological

[58] U.S. GAO, *International Trade: U.S. and India Data on Offshoring Show Significant Differences* (October 2005). The adjective "significant" is misleading and should have been "dramatic."

differences. The GAO report, however, did not provide quantitative estimates for the five factors, although the largest difference by far is apparently that of the United States not counting trade between Indian companies and their U.S. affiliates, which is the way the large majority of all Indian software and business services companies operate. For example, the GAO report states that on a global basis, for which figures are provided, three quarters of U.S. imports of business services take place between foreign firms and their U.S. affiliates. In sum, the GAO report clarifies the questions about the data but does not provide the answers or recommendations as to how to improve the woefully inadequate collection of basic data for this highest growth Indian export sector, which has considerable political sensitivity in the United States.

*   *   *

These are the basic facts for U.S. bilateral trade with China and India. Trade is growing rapidly with both countries, in both directions. The large and growing trade deficits, especially with China, however, cast a dark cloud over the prospect for U.S. export competitiveness in these markets, and for U.S. commercial interests in broader terms. The growing orientation of Chinese and Indian exports to high technology industries—in the manufacturing sector for China and in the software and other business services sector for India—pose a serious challenge for continued U.S. leadership in technological innovation as well. Other dimensions of the economic relationship also need to be considered, such as the emergence of indigenous MNCs and technological innovation in the two countries, which are the subject of the following chapter. The trade data nevertheless constitute the most detailed, up-to-date picture of what is happening, and the stark figures provided here cry out for a more forceful U.S. policy response to regain export competitiveness, particularly for advanced technology industries.

CHAPTER 4

INDIGENOUS MULTINATIONAL COMPANIES AND TECHNOLOGICAL INNOVATION

The previous two chapters examined several important dimensions of the rapid, technology-driven industrial modernization under way in China and India: education, R&D, foreign direct investment, and trade. Another dimension of more recent origin, but which is central to the longer term economic strategy of both countries, is the development of indigenous multinational companies (MNCs) and technological innovation. This was highlighted in Premier Wen's 2005 statement that independent innovation is the national strategy, and in the Indian 2006 National Strategy for Manufacturing, which calls for Indian industry to become a global player.

There is, in fact, forward movement in both countries in the development of indigenous MNCs and technological innovation. China and India are by far the leaders on both counts within the grouping of newly industrialized or emerging market economies. A progress report, however, is less precise and definitive in this area, for several reasons. There is a considerable time lag, compared with the other dimensions, in the emergence of indigenous MNCs and innovation, and thus a briefer period of observation. Official data are especially sparse and not always reliable, and the definition of innovation, itself, is complex and controversial. The broad lines of the forward movement and the principal challenges being faced in both countries are nevertheless fairly clear.

The presentation begins with a discussion of emerging MNCs and then addresses technological innovation, which, to some extent, overlaps MNC performance. Comparisons between the two countries are made throughout. Suggestions are also offered as to how a more current and systematic assessment could be undertaken, which would be useful for U.S. as well as Chinese and Indian interests. The chapter concludes with a discussion of the importance of the "diaspora connection."

Indigenous MNCs

An MNC is defined as a large company heavily oriented to foreign markets for revenue, involving global marketing and, to varying degrees, production abroad. In many industries, this market globalization requires brand-name recognition for leading-edge, quality products.

Joint ventures and acquisitions, both at home and abroad, are often an important part of corporate strategy. R&D to develop new products and to lower production costs are also increasingly essential for firms to become competitive global players.

Two analytic points of departure for assessing indigenous MNC development in China and India are the distinction between private and public sector MNCs and the role of entrepreneurial spirit for success. On the first point, private sector MNCs receive relatively small financial or other government support and need to be actually or potentially profitable in order to secure market-based financing. Public sector MNCs and hybrid companies, supported in large part by the government, in contrast, can obtain public financing in the absence of market credit worthiness, and can thus have objectives other than profitability. Public sector and hybrid companies play a much larger role among Chinese than Indian MNCs, particularly for overseas investment in the energy and industrial raw materials sectors. Government banking and other financial support and a government management role are also prevalent throughout much of the Chinese manufacturing sector. In India, in sharp contrast, indigenous MNCs are predominantly private sector, usually 100 percent, and the government is more likely to pose financial obstacles than to bestow support.

The critical role of entrepreneurial spirit, the second point of departure, is characterized by risk taking and flexibility, the capacity to constantly adjust to changing market conditions and opportunities. All new investments involve risk and some will fail. The entrepreneurial spirit does not punish occasional failure, but rewards diligent risk assessment and a profitable overall batting average. The global entrepreneur needs to act with vision, boldness, and the ability to make constant adjustments in strategy and execution. At this juncture, the entrepreneurial spirit is more deeply embedded and forcefully exercised in Indian than in Chinese MNCs. Chinese corporate management tends to be more risk averse and slower moving in its decisions, often as a result of the public sector engagement, and in part a reflection of the current sociopolitical culture in China.

The Number and Sectoral Scope of MNCs

Both Chinese and Indian MNCs are growing rapidly in number and sectoral scope. China has had a five to ten year head start in manufacturing, but India has recently been closing the gap in some industries and is clearly in the lead for software and business services. In 2006, The Boston Consulting Group conducted a survey in which the top 100 companies in the group of newly industrialized countries, called rapidly developing economies, were selected as being large and increasingly globalized in their market orientation.[59] China and India were the two lead countries by far, with 44 and 21 companies, respectively, followed by Brazil with 12, Russia with 7, and Mexico with 6. By sector, the Chinese companies were most concentrated in the IT and telecommunications hardware (6), consumer electronics (6), automotive (6), home appliances (5), and fossil fuels (4) sectors. Indian companies were most concentrated in the IT and engineering services (6), automotive (5), and pharmaceuticals (3) sectors. The largest overlap was in the automotive sector, while no Indian company made the cut in the IT and telecommunications hardware and home appliances sectors, and no Chinese company was listed for the IT and engineering services and pharmaceuticals sectors. It is noteworthy that two Chinese and one Indian companies were listed for the steel sector.

Outward FDI

Outward foreign direct investment (FDI) by Chinese and Indian firms can play an important role in developing global markets and establishing brand recognition. For private companies, it can establish a local presence in major foreign markets, deepen the capability for technological innovation, and produce the most rapid growth in revenues—the "external growth" from mergers and acquisitions. For public companies, FDI can, in addition, serve various political objectives, including the securing of overseas supplies of fossil fuels and industrial raw materials.

Outward FDI has risen sharply for both China and India beginning in about 2005, although with differing characteristics in the two countries. Chinese FDI has been heavily concentrated in public and hybrid company investment in fossil fuels and raw materials, whereas private sector investment has been predominantly in the manufacturing sector and limited largely to extending abroad the production and marketing of existing product lines. Indian outward FDI, in contrast, has been predominantly private sector, broader in industrial scope, and more venturesome in seeking out new areas of business activity. This contrast reflects both the deeper pockets of Chinese public companies and the more dynamic, entrepreneurial leadership of Indian firms.

Detailed information on outward FDI is limited and can be misleading. Total official figures for 2004-2006 are shown in Table 4-1. Chinese outflows increased sharply from $5.4 billion in 2004 to $12.3

---

[59] The Boston Consulting Group, Inc., *The New Global Challenge: How 100 Top Companies from Rapidly Developing Economies Are Changing the World* (May 2006).

billion in 2005 to $16.1 billion in 2006. The Indian figures were only $2.2 billion and $2.5 billion in 2004-2005, but then surged to $9.7 billion in 2006. These figures for both countries probably understate the actual foreign investment, however, in part because investments financed abroad are not fully reported. This underreporting was especially large for India, when, as explained below, actual investment in 2005 surged to about $8 billion compared with the $2.5 billion official figure in Table 4-1. Further detail, as available, is presented first for China and then for India.

Table 4-1
FDI Outflows
($ billions)

	2004	**2005**	**2006**
China	+5.4	+12.3	+16.1
India	+2.2	+2.5	+9.7

Source: United Nations Conference on Trade and Development, *World Investment Report, 2007*, Annex Table B.1.

China.—Much if not most Chinese outbound investment since 2005, when the sharp rise began, has been in the fossil fuels and raw materials sectors in developing countries—in Africa, the Middle East, Central Asia, and Latin America. It is heavily financed by the government through public and hybrid companies, and the purpose is to secure supply of these resources, with profitability not a decisive factor. In view of ongoing Chinese initiatives throughout these regions, and $1.5 trillion of foreign exchange holdings in the Chinese Central Bank, such investment is likely to continue growing.

The outlook for FDI by Chinese manufacturing companies is less clear. Many companies now have billion dollar annual revenues, largely in exports, and state the intent to invest abroad, but the experience to date is not definitive. Initial investment has been heavily in Southeast Asia, while profitable investment in Europe and the United States has been elusive. Three examples of investment in Europe and the United States have received a mixed assessment at best. The TCL Corporation consumer electronics group acquired the Thomson color television business in Europe in 2004, which made it the largest color television producer in the world, but the performance of Thomson has been disappointing. Huawei Technologies is becoming a major global producer of telecommunications equipment, with a large R&D program spread in many countries. Its principal investment initiative in the

United States, however, a 2003 joint venture with 3Com, failed, with 3Com Corporation buying out Huawei in November 2006, although in late 2007 Huawei attempted to buy back into 3Com. In February 2007, Huawei entered into a joint venture with Nortel in Canada, but again with, as yet, unclear results. The most closely watched Chinese investment has been the Lenovo acquisition of the IBM PC division in 2005, for $1.25 billion, making Lenovo the third largest global PC producer after Dell and HP. The merger, including a complicated management structure, however, has moved forward slowly, and longer term profitability and the maintenance of U.S. and European market shares remain questionable in the face of strong competition from U.S., Korean, and Taiwanese companies.

A recent development that could lead to greatly increased outward FDI was the December 2006 announcement that the China Development Bank (CDB) will expand its financial support for overseas Chinese enterprises.[60] CDB governor Chen Yuan stated that the bank was seeking new international financing opportunities in energy and metals, and that lending for foreign projects had grown very fast and was expected to rise further. "We follow the biggest market players in China [abroad]," he said. CDB has already extended credit to Huawei, Lenovo, Chery, Chinese National Petroleum Corporation, Sinopec, and Minmetals and has sent teams to Venezuela, Russia, and Central Asia. Yuan confirmed that the bank is setting up a $6 billion joint fund in Venezuela, of which China will contribute $4 billion, to build housing, roads, railways, and telecommunications. Such CDB and other public sector financing of FDI raises a number of policy issues related to investment subsidies, which are discussed in Chapter 9.

One vulnerability of the Chinese investors in the cases discussed here is a slower moving and less agile decision making process in the face of unanticipated market developments. It is basically the issue of relatively weak entrepreneurial spirit. For Huawei Technologies, an embedded government connection, particularly with defense industry and the Chinese military, plays a role. The Chinese part of the Lenovo management has been characterized as deeply hierarchical, based on deferential behavior, while top level decision making is shared by a hands-on Chinese chairman and an American CEO, feeling their way forward from diverse cultural settings. Nothing could be more different from the bold and dynamic individuals that lead many Indian MNCs.

India.—The surge since 2005 in Indian outward investment is wide ranging by sector and predominantly private sector in its orientation.

---

[60] The information in this paragraph is drawn from Andrew Yeh, "China Development Bank backs push to invest overseas," www.ft.com, December 6, 2006.

The IT and business services sector led the way, and is now being followed throughout the manufacturing sector and in some service industries. A survey of 600 Indian companies revealed $7.1 billion of outward FDI in 2005, which probably means about $8 billion in total FDI, compared with only $2.5 billion officially reported.[61] The largest industries in the survey were pharmaceuticals ($1.58 billion), banking ($1.18 billion), IT services ($786 million), metals ($778 million), energy ($631 million), autos and parts ($364 million), telecom ($308 million), chemicals ($235 million), fertilizers ($153 million), and petrochemicals $119 million). Among the top 20 destinations by country, $2.9 billion went to Europe, $1.2 billion to the United States, $487 million to Thailand, $376 million to China, $192 million to Singapore, and $187 million to Australia.

Within sectors, numerous companies recorded substantial outward investment. In the pharmaceutical sector, the leaders were Dr. Reddy's ($777 million), Ranbaxy ($324 million), and Matrix ($235 million), followed by seven other companies ranging from $46 million to $17 million. In the IT services sector, the leaders were TCS ($207 million), Wipro ($154 million), and Scandent ($120 million), followed by seven others ranging from $63 million to $11 million. In the automotive sector, the three largest vehicle producers were Tata Motors ($148 million), TVS Motor Company ($27 million), and Mahindra & Mahindra ($27 million), and the three largest auto parts producers were Bharat Forge ($104 million), Ucal Fuel Systems ($23 million), and Sundaran Fasteners ($14 million). The $1.18 billion in the banking sector was almost all accounted for by the State Bank of India's investments in Mauritius, Indonesia, and Kenya.

The surging Indian outward FDI has continued its upward trajectory in many sectors, and an updated survey of CRISIL for financial year 2006-2007 would be revealing. The most ambitious recent outward investments have been in the metals sector.[62] In February 2007, Tata Steel acquired Anglo-Dutch Corus for $12 billion to become the world's fifth largest producer. Also in February 2007, Essan, India's fourth largest steel maker, formed a joint venture with two state run Vietnamese companies to build a $527 million steel mill.

---

[61] CRISIL Centre for Economic Research, *Creating the Indian MNC*, September 2006. The figures are for the financial year from April 2005 through March 2006, referred to here as 2005.

[62] In June 2006, Mittal Steel, led by Lakshmi Mittal, acquired Luxembourg-based Arcelor for $34 billion to become the world's number one steel producer. Mittal is Indian, but he lives in London, works out of corporate headquarters in Luxembourg, and is only now undertaking his first steel investments in India. He is thus a truly international rather than Indian investor.

And yet again in February 2007, Hindalco, part of the Birla Group, announced the acquisition of Novelis, the largest U.S. aluminum producer, for $6 billion, with the intent of creating a globally integrated aluminum producer.

Comparative Performance by Sector

Emerging Chinese and Indian MNCs overlap and compete in a number of sectors, and these relationships will intensify over time and become an important dimension of the joint rise of the two nations to become global economic powers. There is little available sectoral analysis of comparative performance, however, and such studies would be useful for understanding not only the evolving Chinese-Indian relationship, but for the impact their growing joint engagement will have on the global composition of particular sectors. Brief anecdotal comments are offered here for the principal sectors as a stimulus for such sectoral analysis.

1. *IT and telecommunications.* Chinese companies are far ahead as global players for IT hardware, while Indian companies are fully established as leading-edge MNCs for IT software and business services. The sectoral engagement is broadening in both countries, however, to include both hardware and software, while relationships with rapidly growing domestic telecommunications companies are deepening. Chinese MNCs, some linked to Taiwanese firms, are thus far principally domestic market-oriented, but their strategy is to extend to overseas investment, including R&D, as in the cases of Lenovo and Huawei. Indian hardware and telecommunications companies remain predominantly oriented to the high-growth domestic market, but they, too, have global aspirations.

2. *Consumer electronics and appliances.* China is far out front, with large, export-oriented companies seeking to establish quality, brand name reputations, such as Haier for white goods home appliances, including in the U.S. and European markets. Indian companies are beginning to mobilize and bear watching, but they have a way to go before becoming influential MNCs.

3. *Pharmaceuticals.* Both Indian and Chinese companies have large production and exports of generic drugs and are building their R&D programs to eventually develop their own patented drugs. Indian companies, led by Ranbaxy and Dr. Reddy's, are currently ahead of their Chinese counterparts, including through FDI and R&D programs in Europe and the United States.

4. *Automotive vehicles and parts.* This sector could become the most intensely competitive as both nations pursue similar strategies.

Vehicle and parts production initially centered on the large and rapidly growing domestic markets, with exports progressively developed for auto parts on a global basis and for vehicles for export initially to developing regions such as Southeast Asia and the Middle East. Major vehicle exports to Europe and the United States are now planned and will be the decisive test for becoming fully competitive global companies. Chinese vehicle producers, such as Chery and Shanghai Automotive Industry, were built up as joint ventures with foreign companies and are now striking out on their own. Indian companies, after failed joint ventures as junior partners in the 1980s and 1990s, have recently entered joint ventures on a more balanced basis, such as Tata Motors with Fiat in 2005 and Mahindra & Mahindra with Renault-Nissan and International Truck and Engine Corporation of the United States in 2006.[63] Indian auto parts producers are behind their Chinese counterparts in exports, with $9 billion Chinese versus $2 billion Indian in 2006, but have a number of potential global competitors, including at least 12 companies that have received the Japanese Deming award for quality. Bharat Forge has the world's largest engine and axle forging factory, with sales projected to rise to $1 billion by 2010. An overriding question about international competitiveness in the automotive sector is the size and performance of R&D programs, which can be very large. Chinese vehicle companies are expanding internal R&D rapidly, as part of their drive to become independent producers. Indian companies, in contrast, tend to rely heavily on their foreign partners for R&D financing, while offering low-cost engineering design and manufacturing as their part of longer term joint venture relationships.

5. *Ferrous and nonferrous metals.* The steel sector, facing severe global overcapacity over the next several years, will be greatly influenced by Chinese and Indian firms, but in different ways. Chinese domestic production will be the source of much of the excess supply, with growing exports at low prices likely to result in an increase in anti-dumping actions, while the rise of Tata/Corus and other Indian steel companies located in importing countries could put them on the other side in responding to the anticipated surge in Chinese exports. The course ahead for Chinese FDI in nonferrous metals in developing countries and Indian engagement, for example through the acquisition of Novelis by Hindalco, is less clear, and the two country situations will probably be less closely related than it is for steel.

6. *Fossil fuels.* Companies in both countries will play growing roles in this sector through increased domestic production and outbound FDI, but again in different ways, and with China much larger in quantitative

---

[63] See "The New Laws of Attraction," *Business World*, January 2007, pp. 28-33.

terms. Chinese FDI in petroleum in developing regions has already been noted, and the failed attempt of Cnooc to acquire Unocal in the United States in 2005 was an indication of global aspirations in the industrialized regions. For coal, China Shenhua Energy is a $10 billion firm that is seeking to become a diversified global mining venture, with acquisitions under discussion in Mongolia, Vietnam, and Australia. Indian companies, for the most part, are less engaged globally, although Reliance Industries is a $25 billion integrated company, ranging from oil and gas production to refining to petrochemicals and manmade textile products. It has the world's third largest oil refinery refining very heavy crude for reexport, and will almost double its refining capacity by 2008 to have the world's largest refining complex.

* * *

In conclusion, a rapid development of MNCs is under way in both China and India, which will further strengthen their already predominant role within the newly industrialized country grouping. The combination of a very large domestic market, ample engineering and other human resources, and various policy incentives is the basic mix of ingredients for the momentum in a number of sectors.

China is ahead in overall scale and scope, particularly in its much larger base of manufacturing industry and exports. India, however, is well ahead for IT and other business services, and to a lesser extent for pharmaceuticals, and has major engagements in the steel and automotive sectors. The overlap in sectoral engagement and direct competition between the two nations is thus far relatively limited and is related in large part to the two factors introduced as points of departure. The Chinese government provides far larger financial support to and management participation in both public and private sector Chinese companies, which will intensify as the recently launched policy of indigenous enterprise and innovation is implemented. Indian companies, in contrast, are far more independent in their operations.

The second point of departure, entrepreneurial spirit, favors India up to this point, for a number of reasons already noted. One other aspect of entrepreneurial capability of particular relevance to the current outlook is the ability to develop internationally-oriented conglomerate groupings, which have a long history in India but have not yet emerged in China. Three such Indian conglomerates, two very old and the other more recent, make the point. All three have annual revenues of $20 billion-$25 billion, with annual growth in the range of 10 to 30 percent. The Tata Group dates back to the 19th century, with 70 percent of revenues spread among IT services (TCS), steel, and automobiles, and the remainder including energy, chemicals, hotels, financial services, and tea. The Aditya Birla Group, also with 19th century origins, has 74 manufacturing units and sectoral services in 13 countries, including in aluminum, copper, textiles, chemicals, telecom, retail sales, and financial services. Reliance Industries has been engaged throughout the petroleum and petrochemical sectors, as noted earlier, and then in 2006 launched major initiatives to develop the full infrastructure for two town-size SEZs and to create Reliance Retail, a large network of retail supermarkets, recently permitted for private Indian companies. Each of the three conglomerates, moreover, is distinguished by bold, forward-thinking entrepreneurial leaders: Ratan Tata, Kumar Birla, and Mukesh Ambani.

Indigenous Technological Innovation

Technological innovation is a central objective for the longer term trajectory of China and India toward becoming advanced technology superstates. Measuring technological innovation, however, is the most elusive dimension of the overall course of industrial modernization, especially "indigenous" innovation, that is by Chinese and Indian companies or government bodies rather than by foreign companies. It is the most lagging indicator of performance, with relatively short periods of observation in most industries. The term innovation, itself, is variously defined, ranging from new product development to changes in existing products to improvements in the process for producing goods and services. There is also an important overlap between technological innovation for commercial products and for weapons modernization, with the latter developments generally kept secret, which has particular relevance for the U.S.-China relationship. Finally, there is surprisingly little systematic assessment of new technology development and application in China and India, in what are often fast-moving circumstances.

There is no doubt that technological innovation is happening at an accelerating pace in both countries. All of the basic ingredients are engaged on a large scale: R&D expenditures, government finance and other incentives, large multinational high-tech companies, and burgeoning start-up companies financed by venture capital. The questions remain, however, as to how fast and how far the innovation is proceeding and to what extent it is indigenous.

In most areas, China is well ahead of India in terms of resource commitments, private sector engagement, and defense-related innovation. Innovation in India is becoming more prominent, although this appears to be largely if not principally a result of large R&D programs by foreign companies, which highlights a difference between Indian and Chinese policies at this stage. India welcomes without reservation foreign R&D programs as a catalyst for broadening clusters of technological innovation and development. China, in contrast, now places high priority on fostering innovation by Chinese companies, which can be at the expense of foreign companies, and in any event makes foreign companies wary of exposing their most advanced technologies to Chinese counterparts and even to their own Chinese engineers who may later be induced to shift to a Chinese competitor. The net result is a growing attraction of India as a place where foreign companies can locate advanced innovation activities with greater security. An example is the experience of Cisco, which had to challenge Huawei over patent infringements in Chinese courts. Cisco finally won the case, but then, in January 2007, as recounted earlier, opted for India as the principal overseas location for integrating its global R&D and other innovation work.

The presentation here begins with the two standard proxy measures of innovation, patents and articles in professional journals. These measures have shortcomings, but they are useful indirect measures of the rate of growth for innovation within China and India and of change in relative performance between the two countries and the advanced industrialized centers of innovation—the United States, Europe, and Japan. The concluding section provides commentary and anecdotal information about innovation in key sectors. As for questions of definition, some distinctions are made between new products and applied technology for existing products, but no comprehensive definition is offered.

Patents

Patents are a proxy measure of potential technological innovation, but not a direct measure of applied new technologies. Corporate strategies with respect to patent applications vary greatly. Japanese and, increasingly, American companies tend to request wide-ranging patents to stake out intellectual property rights in areas where the actual application of new technology can be highly doubtful or well into the future. Nevertheless, the relative levels and trends of patent applications among countries do provide an indication of changes in innovation performance, with the recent strong upward trends in China and India particularly noteworthy.

Table 4-2 presents patent applications for seven countries for 1995, 2000, and 2005, broken down between resident and nonresident applicants. For 2005, Japan was number one with 427,000 applications, followed by the United States with 391,000. The most striking development shown by the table is the rapid rise of China to third place, with 173,000 applications in 2005, almost ten times the 19,000 in 1995. Germany, the United Kingdom, and France, follow with 60,000, 30,000, and 17,000, respectively. Thus, patent applications in China in 2005 were larger than in the three principal European countries combined. The rise in patent applications in India is less dramatic than that in China, but still a more than doubling from 7,000 in 1995 to 17,000 in 2004.

Table 4-2
Patent Applications
(thousands)

	1995	**2000**	**2005**
Japan, Total	369	420	427
Resident	334	384	359
Nonresident		35	68
United States, Total	228	296	391
Resident	124	162	203
Nonresident	104	134	188
China, Total*	19	68	173
Resident	10	25	93
Nonresident	9	42	80
Germany, Total	46	62	60
Resident	38	51	48
Nonresident	8	11	12
United Kingdom, Total	28	33	28
Resident	19	22	17
Nonresident	9	11	11
France	16	17	17**
Resident	12	14	14
Nonresident	3	3	3
India, Total	7	9	17**
Resident	2	4	7
Nonresident	5	4	10

*China mainland only. Patent applications in Hong Kong in 2005 were 10,000, almost all nonresident.
**2004.
Source: World Intellectual Property Organization, *WIPO Patent Report: Statistics on Worldwide Patent Activity*, 2007 edition.

The breakdown between resident and nonresident applications in 2005 is about the same for the United States and China, at slightly more than half resident. The resident share is much higher in Japan and the European countries, however, while substantially less than half for India. The one measure that would be of special interest but is not included in the data is the relative shares of resident applications by foreign as compared with domestic Chinese and Indian firms. Resident applications by foreign firms imply innovation under way within China and India, but not by domestic firms. Perhaps this information can be made available.

Table 4-3 presents patents granted in the United States for the same years and for the same countries. Clearly China and India are far behind the others, which indicates relatively small penetration of the U.S. market for goods newly patented at home. Both countries do, however, show a sharp rise in patents granted, more than five times larger for China in 2006 compared with 1995 and more than ten times larger for India. In 2006, the Chinese level of 1,723 was close to one-half the U.K. and French levels, and on current course Chinese patents would overtake U.K. and French levels within five to ten years. The figures in Table 4-3 for China are for the Mainland plus Hong Kong. Hong Kong had a larger number than the Mainland in 2000—548 versus 162—but the Mainland pulled ahead in 2006 with 970 versus 753 for Hong Kong. A final comment about Table 4-3 with respect to China and India is that patents granted in the United States were probably predominantly by Chinese and Indian firms, since other multinational companies would likely have filed from their home base or through their U.S. subsidiaries.

Table 4-3
U.S. Patents Granted

Country of Origin	**1995**	**2000**	**2006**
United States	64,510	97,011	102,267
Japan	22,871	32,922	39,411
Germany	6,874	10,824	10,889
United Kingdom	2,685	4,092	4,329
France	3,010	4,173	3,431
China, Total	311	710	1,723
Mainland	63	162	970
Hong Kong	248	548	753
India	38	131	506
Global Total	113,955	176,083	196,436

Source: U.S. Patent and Trademark Office, *Patent Counts by Country/State and Year*, March 2007.

The overall picture for patents is a rapid rise for China and India, both for applications at home and for patents granted in the United States, which reinforces a similar upward trend in innovation. Patent applications in China are approaching the European level, while India still lags well behind the other six listed countries. The missing measure, not contained in available data, is the share of resident patent applications by indigenous companies in China and India, which would be a more targeted measure of indigenous innovation.

Professional Journal Articles

The number of articles in the internationally recognized and peer-reviewed set of science and engineering journals covered by the *Science Citation Index* and the *Social Sciences Citation Index* is another proxy measure for technological innovation. The number of such articles published in the same seven countries for 1995, 2000, and 2003, are presented in Table 4-4. In 2003, the United States was first with 211,000 articles, followed by Japan with 60,000, and the three European nations with 48,000, 44,000, and 32,000. China and India had lower levels, with 29,000 and 13,000, respectively, but also the highest rate of growth over the eight-year period. China had the highest growth by far, at 222 percent, compared with 30 percent for India, 25 percent for Japan, and only 4 percent for the United States. As a result, China was close to the French level in 2003, and on current course will overtake each of the three Europeans by 2010. It is also noteworthy that China was at par with India in 1995, while rising to more than double the Indian level in 2003. This reflects the far greater resources provided by China to university and other public sector

Table 4-4
Science and Engineering Articles
(thousands)

	1995	2000	2003	Percent Change 1995-2003
United States	203	196	211	+4
Japan	48	55	60	+25
United Kingdom	46	49	48	+4
Germany	38	43	44	+16
France	29	31	32	+10
China	9	18	29	+222
India	10	10	13	+30
World	581	632	699	+20

Source: National Science Foundation, *Science and Engineering Indicators 2006*, Appendix Table 5-41.

research since 1995, and reverses the relationship in the 1980s when the Indian output of articles was far higher than that of China.[64]

Table 4-5 breaks down the articles in 2003 by field. The largest share for the United States and the Europeans was in clinical medicine and bioresearch, ranging from 40.7 to 47.5 percent of the total. China and India, in contrast, had the highest shares for biology, chemistry, and physics, at 53.9 percent and 51.6 percent, respectively, compared with 22.9 percent for the United States. China is also far ahead for engineering and technology articles, at 16.8 percent of the total, compared with 7.0 percent for the United States. Japan, like China, has relatively high shares in biology, chemistry, physics, and engineering and technology. In mathematics, the United States and the United Kingdom had the highest shares, at 3.0 percent and 3.1 percent, respectively, compared with only 0.4 percent for China, which is somewhat surprising in view of the prominent role of ethnic Chinese in American university math departments.

Table 4-5

Science and Engineering Articles by Field

(2003, percent)

	Clinical Medicine Bio Research	**Biology Chemistry Physics**	**Math**	**Engineering Technology**	**Psychology Social Services**	**Other**
United States	47.5	22.9	3.7	7.0	6.4	12.5
Japan	40.5	41.8	0.4	12.1	1.8	3.4
United Kingdom	46.3	23.7	3.1	7.1	7.7	12.3
Germany	45.0	34.5	2.0	7.7	4.1	6.7
France	40.7	35.7	1.0	8.6	6.5	7.5
China	18.9	53.9	0.4	16.8	4.4	5.6
India	28.4	51.6	0.2	11.9	3.3	4.6

Source: National Science Foundation, *Science and Engineering Indicators 2006*, Appendix Table 5-45.

[64] See Ronald N. Kostoff, *et al.*, *Assessment of India's Research Literature*, Defense (Technical Information Centre Report, No. ADA 444625, 2006). This bibliometrics report contains a number of other measures as well as copious analysis of Indian and Chinese science and engineering articles. The conclusions are consistent with those drawn here.

These sharp differences in article concentration by field reflect differing political priorities for basic and applied research, with implications for the directions of technological innovation. In China and India, innovation in advanced technology industries for commercial application has top priority, reflected in the research concentration in the natural sciences and engineering. For the United States, in contrast, public spending on health care has the broadest political support, while research related to other commercially-oriented industries receives a lower priority. This issue is addressed in Chapter 10 as an important component of the U.S. domestic policy agenda for responding to the international challenge to U.S. technological leadership.

Innovation by Key Sector

The degree to which innovation is occurring in China and India, and its specific content, can best be addressed sector by sector, where various steps in the innovation process, from idea to final product, can be identified and assessed by industry participants and experts. Unfortunately, relatively little systematic study is available along sectoral lines, and the following commentary provides only the broad lines and a few examples of innovation under way. Even these snapshot observations, however, demonstrate the scope and depth of potential technological innovation in the two countries, and it is hoped will serve as a catalyst for more in-depth, sector-specific analysis. It begins with the aerospace sector, where dramatic developments are unfolding in both countries. The next two sectors are those expected to play major roles in terms of innovation in coming years: nanotechnology and biotechnology. The final three sectors are those most commercially engaged, where large R&D programs and technological innovation play decisive roles: the information technology and telecommunications (ITT), the automotive, and the medical science and pharmaceutical sectors.

Aerospace.—Both China and India are devoting large resources to develop advanced technology aerospace systems, with extensive overlap between commercial and military objectives. China plans to launch more than 30 commercial satellites to form the Beidou global earth observation system, called Compass in English, with initial coverage of China and parts of Asia by 2008. This system would become a competitor to the U.S. GPS system, and its early launch schedule is causing the Europeans to reconsider the development of their Galileo system, which is experiencing delays and rising costs. This Chinese commercial satellite program is moving forward in parallel with its military program, including the January 2007 testing of a direct assault anti-satellite (ASAT) weapon, which was widely criticized as a military space program, and which resulted in space debris that could damage other satellites.[65] China also put its first astronaut in space in 2003, launched its first lunar satellite in 2007, and plans a manned landing on the moon at some future point. Other areas of technological development related to space include military missiles for weapons and, on the civilian side, projects sponsored by the Chinese National Astronomical Observatories (NAO), including the "Giant Eye" telescope which will have the highest spectrum-acquiring rate in the world, the SST spatial telescope for international solar space research, and an infrared vacuum solar tower for solar physics.[66]

The Indian space program is not as advanced as the Chinese program and collaborates more deeply with European and U.S. programs. It is expanding rapidly, however, partly in response to the Chinese program. On the civilian side, the program had earlier focused on satellites for development objectives, such as for communications, remote sensing, and crop production, but it is now being extended to outer-space missions. The Indian Space Research Agency, in cooperation with NASA, will launch an unmanned space mission to the moon in 2008. It plans its first manned flight in 2014 and a manned landing on the moon by 2020, perhaps ahead of China. While joint missions are necessary in the early stages, former space agency chairman Udipi Ramachandra Rao, a key proponent of the lunar program, stated that ". . . we have to develop our own technologies. . . . That is the route the Chinese have taken."[67] The Indian military missile and space programs are also being intensified, again partly in response to Chinese programs. The Chinese antisatellite attack was reported as having "sent alarm bells ringing in India's defense and security establishment."[68] The January 2007 launch of the Cartosat-II satellite brought to near completion the Indian surveillance and reconnaissance program to track troop movements, missile silos, and military installations in neighboring countries.

---

[65] Indian Foreign Minister Pranab Mukherjee warned that the Chinese action "might trigger a new arms race, which could have a negative impact on the defense and development of most nation states." See reuters.com, March 7, 2007.

[66] These NAO projects were cited in Michael Pillsbury, "China's Progress in Technological Competitiveness: The Need for a New Assessment," (Paper presented to the U.S.-China Economic and Security Review Commission, in Palo Alto, CA, April 21, 2005), p. 5.

[67] See K. S. Jayaraman, "India's Space Agency Proposes Manned Space Flight Program," *Space News*, November 10, 2006.

[68] See Rajat Pandit, "China Missile Worries India," IST Times News Network, January 20, 2007.

Looking ahead, if the manned space programs move forward as planned, China and India will join the United States and Russia as the only nations to have the technological capability to launch humans into space.

Nanotechnology.—In 2005, a U.S. Task Force report, "The Knowledge Economy: Is the United States Losing Its Competitive Edge?" concluded that China "has been investing heavily in nanotechnology and already leads the U.S. in some areas."[69] Subsequent reporting has confirmed this assessment. India, in contrast, has lagged far behind in nanotechnology but is beginning to become more engaged.

The Chinese Academy of Sciences knowledge innovation program, launched in 2001, gave high priority to nanotechnology.[70] The program currently includes at least 20 academic institutions, 1,000 to 1,200 scientists as principal investigators, and another 2,000 graduate students as assistants in nanotechnology research. The Shanghai Nanotechnology Promotion Center spent $126 million from 2001 to 2006 to train 1,500 scientists and engineers on the use of specialized equipment and machinery. Government financial and tax incentives promote technology transfer from labs to firms, with a heavy focus on small, start-up companies. In Shanghai alone, there are an estimated 100 to 200 small- and medium-sized companies doing nano-related work. Chinese nano-related publications in scientific journals are second only to the United States, and growing rapidly. China also plays a lead role in commercial application, with more than 30 product lines using nanomaterials, including for textiles, plastics, porcelains, lubricants, and rubbers. China is already a world leader in nanomaterials application, such as coatings and composites.

India lagged far behind in nanotechnology because of scarce public funding and a lack of interest in the private sector, but this is beginning to change. In 2002, India spent only $4 million on nanoscience, compared with $200 million in China. In 2004, however, Indian president A.P.J. Abdul Kalam, a revered scientist, organized a meeting of nanoscience experts to develop the Nano Science and Technology

---

[69] See www.futureofinnovation.org, February 16, 2005. The Task Force consisted of 21 academic and private sector organizations and companies, including the American Electronics Association, the American Physical Society, the Materials Research Society, Intel, Lucent, and Microsoft.

[70] All of the material in this paragraph is from Richard P. Applebaum and Rachel A. Parker, "Innovation or Imitation? China's Bid to Become a Global Leader in Nanotechnology," University of California at Santa Barbara, January 2007. This is a rare example of a thorough sectoral assessment of technological innovation in China and a model for what should be done for other sectors.

Initiative, to be funded at about $40 million a year. In parallel, a 2006 private sector initiative proposed raising $550 million for a large SEZ, Project Nano City, intended to attract billions of dollars of investment in pharmaceuticals, computer software, energy, and nanotechnology. The initiative was taken by California-based Sabeer Bhatia, cofounder of the U.S. company Hotmail, which was sold to Microsoft for $400 million in 1997, and who is now turning his managerial and innovative abilities to his homeland. The SEZ proposal is still in the approval process, however, and the concrete results for nanotechnology development in India remain to be seen.

Biotech.—The same 2005 U.S. Task Force report concluded that China "is making rapid progress in biotechnology." China has been ahead of India for biotech in most respects, but the gap is narrowing as Indian private investment expands. Indian biotech industry sales rose from $1 billion in 2004 to $2 billion in 2006 and are projected to reach $5 billion in 2010.[71] The two dominant sectors involved are agriculture, principally genetically modified (GM) crops, and medical science. Agriculture is addressed here, and medical science included with the pharmaceutical sector below. China spent about $400 million in agrobiotech activities in 2007, and this could double by 2010. The government employs 2,000 scientists in 200 research labs. GM crops have been approved for tomatoes, sweet pepper, and papaya and are being field tested for rice, maize, wheat, cabbage, cauliflower, soy beans, cotton, melon, and tobacco. GM crops have thus far been directed to the domestic market, but the demand for food is expected to soar throughout Asia in the coming 10 to 15 years, and Chinese minister of science and technology Xu Guanhua predicted that agrobiotech "could become the fastest growing industry in China in the next 15 years."[72]

India has a longer-standing program of agrobiotech research, although the $80 million funding in 2007 was well below that of China. Joint research is under way with Cornell University, with support from the U.S. Agency for International Development (USAID), to conduct research on food crops, such as eggplant resistant to shoot borers, potato resistant to blight, and drought- and salt-tolerant rice. India also launched an enhanced biofuel R&D program in 2006, to develop ethanol and jatropha production, with the latter already under large scale cultivation.[73] India is currently ahead of China in the application

---

[71] *Financial Express*, June 8, 2007.

[72] See Salamander Davoudi, "China to quadruple agri-biotech spending," www.ft.com, March 15, 2007. The figures in this paragraph were also taken from this piece.

[73] See *The Economic Times Hyderabad,* December 27, 2006.

of GM crops, as a result of better-established linkages with private farmers. In 2006, India tripled the area planted in biotech cotton alone to reach a total of 3.8 million biotech hectares compared with 3.5 million hectares in China.[74]

Information technology and telecommunications (ITT).—Innovation in this sector has been the central driving force for the technology-driven economic transformation that began in the United States and is now lifting China and India toward advanced technology superstate status. The scope and degree to which ITT innovation is taking place within China and India are wide ranging, complex, and poorly reported. The relative roles of indigenous and foreign company innovation are especially unclear. Principal specific areas for investigation include semiconductor design, high performance computers, telecommunications equipment, and software application. There is no doubt that innovation is taking place in both countries and will grow in intensity. Beyond that, the starting points for investigation are that China is far ahead in innovation related to hardware, India for software, and that interaction between the two economies and with the global process of innovation will deepen.

China has placed top priority on ITT innovation, with significant across-the-board results. Almost all multinational computer and telecommunications companies have R&D centers in China, while Chinese companies are achieving a growing array of results as well. Four hundred and fifty semiconductor design companies have elevated China to second place in this area, still well behind the United States, but progressively closing the gap. The Godson II central processing unit (CPU) computer chip to support the 64 bit Linux operating system was the first high-performance CPU chip for which China has proprietary intellectual property rights. Innovation for computer hardware is also moving up the technology ladder. The Dawning 4000A Shanghai supercomputer can operate at 10 trillion calculations per second, a speed that puts it third behind U.S. and Japanese supercomputers. Chinese military modernization involves a wide range of ITT innovation, including for ballistic missiles, secure communications networks, and cyber warfare programs, increasingly performed by Chinese companies that are also engaged in commercial markets.

---

[74] See Kathryn McConnell, "Asia Seen as Next Focus of Agricultural Biotech Production," www.gov/ei/economic_issues/biotechnology.html, February 16, 2007. The top eight nations in terms of commercial biotech crops in 2006, besides India and China, were the United States (54.6 million hectares), Argentina (18.0 million), Brazil (11.5 million), Canada (6.1 million), Paraguay (2.0 million), and South Africa (1.4 million). India and China are thus in fifth and sixth places.

India is a global leader in software development, which is innovation in advanced technology application, principally as customized application of existing systems rather than as patented new products. The deepening connection between high performance hardware and advanced design software is drawing together the innovation performance in both parts of the ITT sector in both countries.

Automotive.—Chinese and Indian vehicle and parts producers are becoming more engaged in innovation, but R&D expenditures remain far below those of the multinational foreign companies, and it is unclear how much indigenous innovation will emerge. For vehicle producers, new model design can be very expensive, while growing R&D programs in the sector are devoted to the development of hybrid, fuel cell, and electric engines. A wide range of auto parts are also constantly upgraded through more advanced design and new technology application. Both Chinese and Indian companies have large domestic markets and growing exports, but this has not yet led to leading-edge innovation, with rare exceptions. Hybrid and other new engines are under development within Chinese companies, but with little to show for it through 2007. The largest global producer of electric cars is the Indian company Riva, whose production is scheduled to rise from 6,000 vehicles in 2007 to 35,000 in 2009, but this vehicle was developed in Silicon Valley before the company's corporate leader returned home to India for manufacturing. Again, a difference between the two countries is China's greater emphasis on indigenous innovation. China has always limited foreign investment in vehicle production to 49 percent of joint ventures, intending that the Chinese partners would ultimately become independent producers, with a strong capability for innovation. Indian companies, in contrast, as noted earlier, are developing foreign joint ventures with more complementary than potentially rival roles, and whereby the bulk of the R&D and innovation remains with the far larger foreign partner.

Medical science and pharmaceuticals.—Medical science is a priority area for R&D leading to innovation in both China and India. Much research takes place in government laboratories, while pharmaceutical and biotech companies are the driving force for private sector innovation. This is another sector of fast-moving developments worthy of continuing assessment. As of 2007, China is well ahead of India in terms of government-financed research, whereas Indian companies, such as pharmaceutical leaders Ranbaxy and Dr. Reddy's, are ahead of Chinese competitors in developing global markets and innovation.

Chinese Academy of Sciences funding was $1.1 billion in 2005, triple that of 1997, and is scheduled to grow another 70 percent by 2010. Within the Academy program, research on infectious diseases receives high priority in view of political anxiety about the spreading within China of lethal diseases such as acute respiratory syndrome (SARS), bird flu, and HIV. French scientists are assisting to build a laboratory in Wuhan for diseases such as SARS and ebola that could become the most sophisticated in Asia. Chinese pharmaceutical companies have also been expanding R&D and testing capabilities. So far no major drug has been brought to international market, but about 100 innovative drugs are undergoing testing in China, compared with 1,000 in the United States.[75] Much of the innovation work is in collaboration with foreign companies, such as the Swiss firm Novartis, which is building its research facility in Shanghai, which will have 500 scientists, as an integrated facility to perform early drug discovery and clinical tests. Innovation start-up companies abound, and some are engaged in the subcontracting of testing for U.S. companies. Chunlin Chen, CEO of the contract research company Medicilon, states that "a research dollar spent in Shanghai can stretch several times further than it would in the U.S."[76] A major obstacle to innovation by Chinese firms is weakness in regulatory procedures for quality standards, with resulting faulty-product scandals, and delays in product approval, all of which are under review by the government.

Indian government research for medical science, as for agrobiotech, is far smaller than that of China, and the laboratory work in India does not hold the promise predicted for China. Indian pharmaceutical firms, however, are expanding rapidly to become major multinational companies, with global markets for generic drugs expanding to include a few new patented products. R&D expenditures as a share of revenue have risen to 7 percent for Ranbaxy and 14 percent for Dr. Reddy's. Bob Gatte, vice president of Grace Davison Discovery Sciences, compares Indian and Chinese firms: "I don't see that in China there are at the present time companies that are reaching the sophistication of a company like Ranbaxy."[77] Indian biotech companies are also becoming internationally competitive for medical products. Biocon has a distinguished international board and strong management that has filed more than 100 patents for such processes as the manufacture of r-human insulin and a mobile heart monitoring device. Reliance Life Sciences, the biotech unit of Reliance Industries, has 200 international patent filings, and it plans to expand its international pipeline for innovative products through its January 2007 acquisition of U.K.-based

---

[75] See Jean-Francois Tremblay, "China Strides Toward Global Pharma Role," *Chemical & Engineering News* (American Chemical Society, March 12, 2007). This article is another example of a fairly broad sectoral analysis.
[76] *Ibid.*
[77] *Ibid.*

GeneMedix for $35 million. In the other direction, the French firm Merieux Alliance acquired a majority interest in Hyderabad-based Shantha Biotechnics in November 2006, with the intention of making the joint venture a global R&D hub for vaccine research. Shantha was only founded in 1993, developed and marketed a recombinant hepatitis B vaccine in 1997, and has expanded its portfolio to include other vaccines, therapeutic proteins, and monoclonal antibodies.[78]

A Tentative Assessment and the Need for Systematic Sectoral Analysis

The material presented here clearly demonstrates that China and India are both rising toward becoming major advanced technology participants in the global economy through indigenous MNCs and technological innovation. They are by far the two most important countries within the grouping of newly industrialized or emerging market economies, and in some sectors they are already leading-edge innovators, producers, and exporters. The pace of advance up the technology ladder, moreover, is in some areas rapid and dynamic, with multiple market forces and government incentives in play. It is thus a complex task to get a precise fix on where the more advanced technology industries in the two countries are currently, and where they will be in two, five, or ten years' time.

This entire picture is directly relevant to a central question addressed in this study as to whether the United States can maintain its long-standing technological leadership in the various key sectors. There is little question that the leadership gap is narrowing and, in some cases, disappearing, but the scope and pace of the narrowing cannot be assessed here. Any projected assessment, in any event, also needs to take account of the pace of innovation and applied new technologies that will take place in the U.S. economy, inasmuch as all advanced technology boats are rising with the tide of technology-driven innovation and investment.

An important conclusion drawn here, therefore, is that there needs to be more systematic and timely assessment of developments within China and India, as well as greater public awareness in the United States of the challenge posed by China and India to continued U.S. technological leadership. These two conclusions, of course, are linked: the greater the knowledge about developments in China and India, the more likely will be the rise in public awareness necessary to support the appropriate U.S. policy response. Suggestions for more systematic and

---

[78] See *BusinessLine*, November 10, 2006, at www.thehindubusinessline.com/2006/11/10/stories.

timely analysis are offered here, while the issue of greater public awareness as prerequisite for a more effective U.S. policy response is left for later chapters.

Systematic and timely assessment can best be addressed on a sectoral basis by the people who know the sector through private sector engagement in it or close monitoring from a position at a university or consulting firm. A team of such people could produce a reasonably comprehensive assessment of recent developments in China and India, covering innovation, new-product production, and trade, including the distinction between indigenous and foreign company contributions. A first step would be to develop a general sectoral model of questions to be addressed and basic information to be collected. A common study format for all sectors would be helpful for assimilating sectoral assessments into a broader analysis of technology-driven industrial modernization in the two countries, and would be more user friendly for readers trying to understand what is happening. The first such sectoral studies would be the most difficult and time-consuming in terms of organizing systematic investigation, accumulating basic files for data, and establishing public and private sector personal contacts. Periodic follow-up assessments every year or two, given continuity in where and how the assessments are undertaken, would require less time and fewer resources.

There are various places where such sectoral assessments could be developed and managed. A U.S. initiative could be coordinated by the National Academy of Sciences or the Sloan Foundation Industry Studies Program.[79] The lead management role for specific sectors could be given to universities with programs already engaged in studying the particular sector, consulting firms with experience in such sectoral analysis, industry associations, or a team effort among them. Individual participants could include senior corporate managers closest to R&D and product markets in China and India, academic and consultant experts, and government experts from the economic departments and the intelligence agencies.

There should also be a broader international interest in such sectoral analysis, and a multilateral approach could be developed, most readily within the OECD Science and Technology Directorate. Directorate experts would manage the analysis, with the participation of public and private sector expertise from member countries and nonmembers China and India. China has been a nonmember participant in directorate activities, including at the annual ministerial level meeting, for several years but India thus far has not shown an

[79] As a matter of full disclosure, the Sloan Foundation provided financial support for the present study.

interest in participation. Such OECD institutions, however, would probably take considerable time to organize, and would benefit from the experience and results of a faster-off-the-mark U.S. initiative. The suggestion here, therefore, is to begin sectoral assessments within a U.S. structure, while, in parallel, beginning discussion of OECD studies with broader international participation.

These are relatively straightforward suggestions, at modest cost, for producing up-to-date, detailed sectoral assessments that would go a long way to understanding not only the enormous mutual economic gains from trade and investment through the emergence of indigenous MNCs and innovation in China and India, but also the critical challenge to U.S. policymakers for maintaining U.S. technological leadership. In terms of U.S. technological leadership, the desired outcome would be to increase U.S. public awareness to the point where the U.S. government would place the same top priority on technological innovation that the Chinese and Indian governments do.

Epilogue: The Amazing Diaspora Connection

A recurring theme throughout this study, which is especially important for this chapter about indigenous MNCs and innovation, is the role of the Chinese and Indian diasporas in the United States, including the recent large and growing reflow back home. The diaspora connection is playing a major role in the rapid industrial modernization under way in China and India and is an important and deepening component of the U.S. relationship with both countries. The whole process of outward migration and return flow is an extraordinary story of migration-driven economic development, and there has never been anything like it in terms of quantitative or qualitative impact. Like so many other elements in this story, moreover, the patterns for China and India are parallel, but with key differences as well.

Outward migration and diaspora communities date back centuries for both countries. A large Chinese emigration began in the 15th century, related to the brief period when China was a maritime power, which led to ethnic Chinese communities throughout Asia. A migratory outflow to the United States in the 19th century by impoverished Chinese seeking work, including building railroads, laid the foundation for the Chinese-American diaspora. Indian emigration surged during the British Raj to Africa, the Caribbean, and England. These earlier overseas Chinese and Indians were distinguished by outstanding performance in commerce and family values that place high priority on education.

The diaspora connection related to this study is of more recent origin, the emigration to the United States since the 1950s for India, and somewhat later for China. The process in both cases was in three stages. Stage one was the out-migration to the United States, largely for education, and mostly by members of wealthy and elite families. The American university system was of the highest quality, which was especially appealing to Chinese who could afford it in the wake of the destruction of the Chinese university system during the Cultural Revolution. Indian and Chinese students thrived on U.S. campuses and developed a reputation for high academic achievement, especially in engineering and science.

The second stage was the remaining in America after graduation of a large cumulative inflow of those who had completed their education, together with a later inflow of high technology-skilled immigrant workers. This was the result both of poor job prospects at home and of the rapid growth of U.S. high-tech industry in Silicon Valley and elsewhere. In India under the socialist rule of Nehru and Indira Gandhi there was little professional opportunity in the private sector, and the basic options for college graduates were to secure a position in the government bureaucracy or seek employment abroad. Chinese graduates faced not only low-paying jobs at home but also political oppression culminating in the 1989 Tiananmen Square massacre. By 2000, there were 1.4 million Chinese-born and 1.0 million Indian-born residents in the United States. This large emigration of the most highly educated young people, the "brain drain" phenomenon, was viewed negatively in both countries in terms of economic development. In India, the diaspora was viewed with humor as materialistic and culturally adrift, as in the recent Indian film, *Bride and Prejudice*, while in China, graduates staying abroad were harshly denigrated in political terms. The only positive result for India and China was the remittance payments from the diaspora, which by 2005 had risen to $9 billion to $10 billion a year from the United States to each country.

In the United States, the Indian and Chinese engineers and scientists quickly rose to prominence in the private sector and on university faculties, the Indians in the lead in the private sector, reflecting their more market-oriented culture and English-language capability. From 1995 to 2005, 25 percent of start-up engineering and technology companies in the United States had at least one foreign-born founder, and India was far ahead with 26 percent of the total foreigners, followed by China and the United Kingdom with 7 percent each, and Taiwan with 6 percent.[80] Foreign-born founders were

---

[80] See Vivek Wadhwa, *et al.*, *America's New Immigrant Entrepreneurs* (Duke University, School of Engineering, January 2007), for a detailed account of the share of foreign-born entrepreneurs and innovators, from which the figures here are taken.

especially prominent in California, as participants in 39 percent of all start-up companies, 80 percent of which were in software and innovation/manufacturing-related services. The large Indian role in Silicon Valley, in fact, dates to the 1970s, with such founding innovators as Pallab Chatterjee of Texas Instruments and Tom Kailath at Stanford University.[81] Chinese engineers and scientists were more prominent in research and on university faculties. For innovation, as measured by patent applications, Indians were close to the combined level for China and Taiwan. From 1998 to 2006, foreign born Indians accounted for about 11,000 such applications, compared with about 14,000 by Chinese and Taiwanese.

These management and innovation roles by Indians and Chinese in American companies and universities paved the way for the third stage of the diaspora story, the growing reflow back home, which began in the late 1990s. Rapid growth in advanced technology industries in both countries created relatively high-paying jobs in companies and opportunities to start up new companies. The combination of American university training and job experience put returnees at a premium in the job market, while both governments now welcomed the returnees, offering financial and other incentives. The annual conference for nonresident Indians, or "NRIs," in New Delhi in January 2007, was attended by 1,200 NRIs to hear an optimistic speech by Prime Minister Singh about the new knowledge-based Indian economy, receive achievement awards from President Kalam, and be welcomed by a plethora of state and company recruiting stands throughout the fairgrounds. The annual number returning to China rose from 9,000 in 2000 to 30,000 in 2006, for a cumulative total of about 170,000, including 60,000 in Shanghai alone. Comparable figures are not available for India, but the reflow is likely to be of similar magnitude.

The entire three-stage diaspora experience has been an enormous success story in terms of economic development in the two largest developing countries that together comprise almost half of the developing world. Quality education and work experience in the United States and elsewhere produced hundreds of thousands of entrepreneurs, engineers, and innovators, who are now among the most productive people throughout the economic power structure in both countries. Moreover, the returnees retain close personal and corporate ties with the American diaspora and thus constitute an important linkage for the continued growth in trade and investment across the

---

[81] See Shivanand Kanavi, *Sand to Silicon: The Amazing Story of Digital Technology* (Rupa & Co. 2006), for a highly readable history of the Silicon Valley experience, which highlights the participation of dozens of Indians who played prominent entrepreneurial and innovative roles.

Pacific. And the story continues, with tens of thousands of student emigrants and graduate returnees crossing the Pacific in both directions each year.

The diaspora experience has also had positive political results. The returnees have studied, worked, and lived in the United States, absorbing the American political culture of democracy, rule of law, individual rights, and checks and balances to curb government power and corruption. The returnees carry these values home, become leaders within the economic power structure and attain political influence as well. As a result, the cadre of educated, relatively affluent supporters of democratic values grows, with political impact far greater than the numbers indicate. This pro-democracy impact is positive without qualification for India, whereas in China it increases the challenge to the Chinese government that has yet to define the path to its officially proclaimed objective of democratization.

The diaspora experience also has important results for the United States. In economic terms, the overall impact is for higher economic growth and gains from trade for all three countries, although the growing reflow raises the stakes for the U.S. policy response to restore export competitiveness and maintain technological leadership. In political terms, the pro-democracy effect is again unreservedly positive for the U.S.-India relationship, while intensifying the challenge for the United States in managing its political relationship with China. And finally, and perhaps most importantly, the diaspora experience produces a wide and deep skein of personal relationships between the several million ethnically Chinese- and Indian-Americans and the hundreds of thousands of returnees back home. Both the numbers and the influential roles of the returnees create a diaspora dialogue of great importance for mutual understanding and more positive government-to-government relationships across the Pacific. Add to this, in keeping with the principal theme of this study, the rise of China and India to become affluent, advanced technology superstates, including ubiquitous communication by phone, screen, and email, and higher Asian incomes that make trans-Pacific travel possible, and Rudyard Kipling is in need of revision: The East and West are meeting, with the diaspora connection playing a critical role.

CHAPTER 5

A Net Assessment: Two to Five Years Ahead and Beyond

The previous three chapters have analyzed the extraordinary 8 to 11 percent growth paths of China and India over the past several years, driven by market-oriented economic reforms, open trade and investment, and with a central role for the rapid development of advanced technology industry. Two questions emerge from this analysis. The first pertains to the short to medium term, the coming two to five years. Will the two countries continue the high, 8 to 11 percent growth path? Many forces are in play and many obstacles may impede such continued high growth in both countries, although the obstacles are very different in content. The second and more far-reaching question is whether the two countries, even if the very high growth should falter somewhat over the next several years, are firmly launched on a longer term path toward advanced technology superstate status, as global economic and political powers in the same category as the United States.

As for the first question, the conclusion drawn here is that India is highly likely to maintain an 8 to 10 percent growth path over the next two to five years, whereas China is at least as likely to experience serious adjustment problems that will reduce growth, perhaps to the order of 5 to 7 percent for at least a couple of years. This differential may not appear to have major consequences. China is 10 years ahead of India, in many respects, on the path of industrial modernization, and "only" 5 to 7 percent growth for a few years would be considered a strong economic performance for almost all other countries. The broader implications of these relative trajectories, however, are substantial. For India, continued high growth would confirm sustainable longer term high growth toward advanced technology superstate status, as the third largest economy after the United States and China. For China, slower growth, even for a couple of years, would raise questions about more fundamental political as well as economic change ahead. Moreover, reduced growth in China would have uneven impact, with the most severe adverse effects on the manufacturing sector and the industrialized coastal cities at the core of the economic power structure. Moreover, such a differential over the next several years would change the longer term outlook for the relative power positions of the two countries within the global economic and international political orders.

The net assessment in response to the second, longer term question is that yes, both China and India are now firmly launched toward advanced technology superstate status. The earlier 2005 study had already reached this conclusion for China, and the subsequent two years of high performance in advanced technology industries have validated the earlier assessment. Even if China experiences a somewhat disruptive short- to medium-term adjustment, with lower growth and adverse impact on some industries, the deepening process of industrial modernization and technological achievement will continue. The only caveats to this assessment are a harsh nationalist course that alienates foreign investors and political collapse, both of which are extremely unlikely. India still has a much longer path ahead before achieving superstate status, but the mutually reinforcing economic and political forces in play indicate an accelerating pace of industrial modernization and technological achievement, with reasonably assured ultimate success.

The assessment for the two to five year projection begins with a broad look at the sharp contrast between the structures of the two economies, with the Indian economy experiencing balanced growth, while Chinese growth is excessively export-oriented and headed toward even greater imbalance. This is followed by a discussion for each country of eight major issues that can be policy challenges or economic obstacles for sustaining high growth, and that together will be decisive for the outcomes over the coming two to five years. The concluding section of the chapter then examines more broadly the longer term question of how China and India will continue on the road to become advanced technology superstates.

Growth Paths in Sharp Contrast

The sharp contrast between the balanced growth in India and the increasingly unbalanced growth in China can be demonstrated, in broad macroeconomic terms, by four basic measures, as shown in Figures 5-1 through 5-4.[82] All four figures compare India and China over the period 2002 to 2007, and all four show a striking contrast.

Figure 5-1 shows the current account balance as a percent of GDP, measured on the exchange rate basis. The Indian current account has

---

[82] The discussion of China in this section derives largely from the more comprehensive presentation by Nicholas R. Lardy, "China: Rebalancing Economic Growth," in *The China Balance Sheet in 2007 and Beyond*, C. Fred Bergsten, *et. al.*, eds. (Peterson Institute for International Economics and CSIS, May 2007), pp. 1-24. Lardy concludes that more vigorous policy action by the Chinese government is required, but he does not attempt to project growth.

been close to balance, ranging from 2 percent surplus to 2 percent deficit, whereas the Chinese current account surplus has surged from 2 percent of GDP in 2002 to 9.5 percent in 2006 and a projected 12 percent in 2007. Moreover, in terms of the "basic balance," that is the current account plus long-term capital flow, India is close to balance, with a growing inflow of FDI and other private investment offsetting the current account deficit since 2005, while for China, the large net inflow of FDI adds to the current account surplus, producing a basic balance surplus in 2007 of about 14 percent of GDP. These Chinese surpluses are far beyond any experienced by a major trading nation over the past 60 years, and are a major and growing concern for Chinese trading partners, as described in Chapter 8. In terms of the structural imbalance of the Chinese economy addressed here, it is noteworthy that the current account surplus results in an equal net outflow of savings for investment abroad, principally through Chinese Central Bank purchases of U.S. Treasurys and other financial assets.

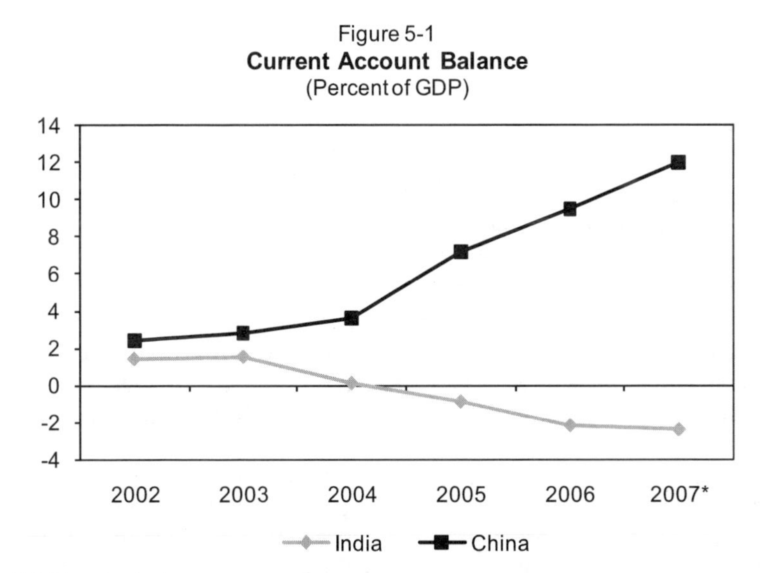

*2007 projected.
Source: International Monetary Fund, *World Economic Outlook Database*, April 2007.

Figure 5-2 shows the level of domestic investment as a percent of GDP, and again there is a sharp contrast. The Indian investment share of GDP has been rising steadily from 23 percent in 2002 to 34 percent by 2006, which for a newly industrialized economy at the early high-growth takeoff stage is about right. The Chinese investment share of investment, in contrast, rose to the extraordinarily high level of 44 percent of GDP in 2006 and an estimated 45 percent in 2007, levels generally viewed as greatly excessive, and an indication that the

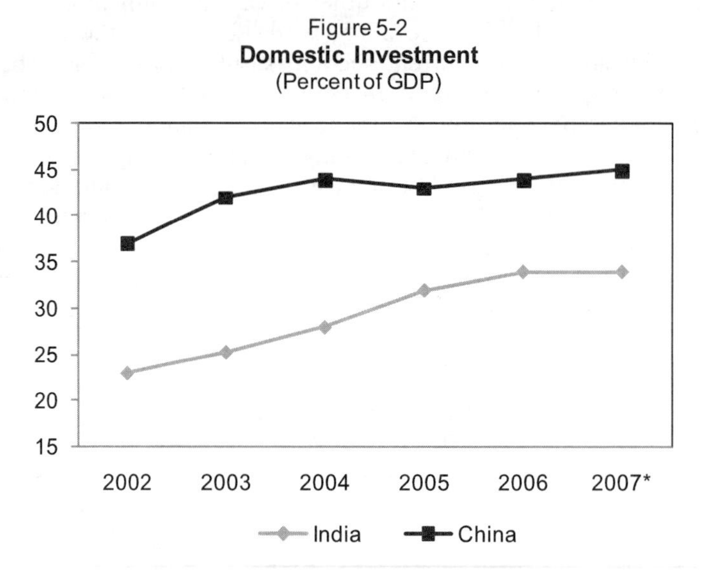

Figure 5-2
Domestic Investment
(Percent of GDP)

* 2006 and 2007 projected.
Source: Ministry of Finance, Government of India, *Economic Survey 2006-2007.* National Bureau of Statistics of China, *China Statistical Yearbook 2007.*

productivity and growth-generating impact of investment in China has become exceedingly low. The common quip in IMF circles is that with investment at 45 percent of GDP, why is Chinese growth **only** 11 percent.

Figure 5-3 presents domestic savings as a percent of GDP and is derived from Figures 5-1 and 5-2, based on the immutable equation that savings equals investment plus the current account balance. Here the contrast between the two countries is even more startling. The Indian savings rate has been rising from 24 percent in 2002 to a projected 32 percent in 2007, roughly keeping up with the rise in investment, and since 2005 the difference has been made up by the current account deficit, which equates to an inflow of savings. The Chinese savings rate, in enormous contrast, has soared from 40 percent in 2002 to 50 percent in 2005 to a projected 57 percent in 2007. This savings rate is in uncharted waters for a large, highly industrialized economy, with consequent domestic consumption down to only 43 percent of GDP.

This extraordinary low level of domestic consumption raises serious questions about the ability to manage a smooth transition from export-oriented to domestically oriented growth.

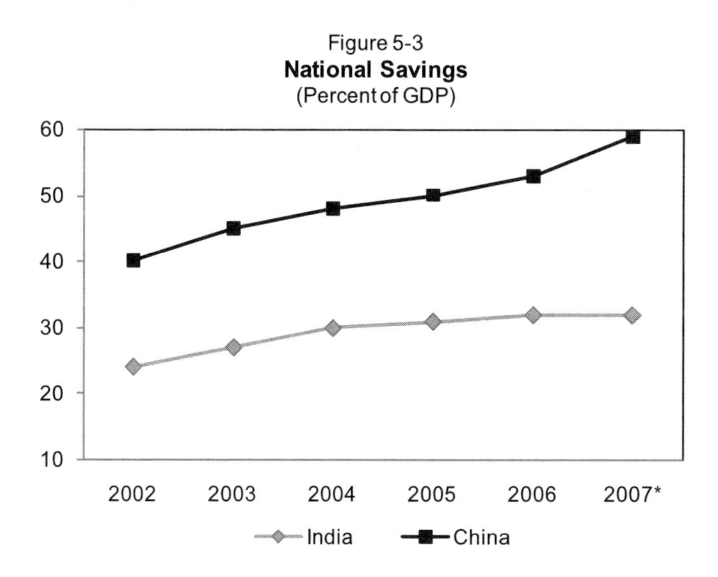

Figure 5-3
National Savings
(Percent of GDP)

* 2007 projected.
Source: Ministry of Finance, Government of India, *Economic Survey 2006-2007.* National Bureau of Statistics of China, *China Statistical Yearbook 2007.*

Figure 5-4 presents personal consumption as a percent of GDP, and yet again the contrast is stark. Indian personal consumption is reasonably high, close to 60 percent of GDP, compared with about 70 percent in the United States, the difference reflecting the need for a higher GDP share of investment in high-growth India. The Chinese personal consumption share of GDP, in striking contrast, is far lower and in steady decline, from 43 percent in 2002 to 39 percent in 2005, and to a projected 36 percent in 2007. This huge difference in personal consumption between India and China—at 60 percent versus 36 percent—is a result both of the much higher rate of savings in China and of the fact that government consumption is a higher share of total consumption in China. The very low level of Chinese private consumption narrows further the market-oriented base for transition from export-led to domestically led growth. It also raises questions of equity for the Chinese people, in that they receive such a small share of economic output for their personal consumption. The resulting comparison of the levels of per capita personal consumption in China and India, on a PPP basis and taking account of the differentials in personal consumption shares of GDP and population size, puts Indian per capita personal consumption at about 80 percent of Chinese. Moreover, with growth far more oriented to personal consumption in India, the difference could narrow further.

Figure 5-4
Personal Consumption
(Percent of GDP)

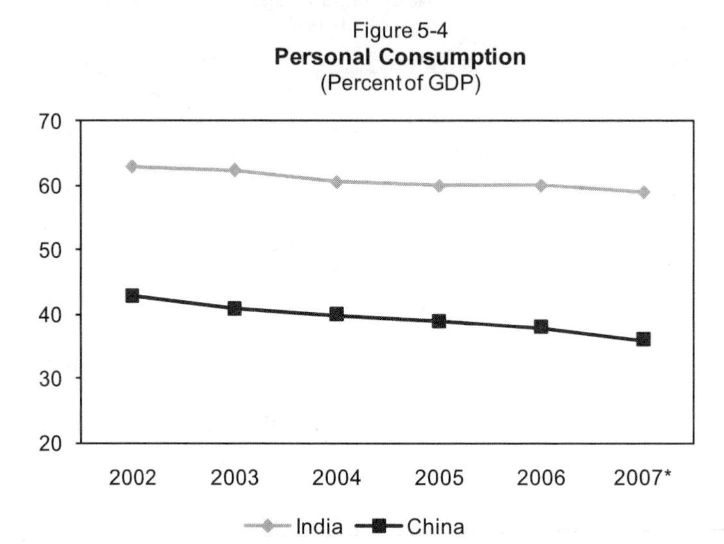

* 2007 projected.
Source: Ministry of Finance, Government of India, *Economic Survey 2006-2007.* National Bureau of Statistics of China, *China Statistical Yearbook 2007*

These are the basic components of the sharply different economic structures that will make it more difficult for China to maintain 8 to 11 percent growth over the coming two to five years. India simply has to maintain its current course of balanced growth, with a relatively high level of investment combined with strong domestic consumption, and with external accounts in reasonable balance. In 2005, Indian investment grew by 16 percent and consumption by 7 percent. China, in final contrast, faces formidable policy challenges and economic obstacles to restructuring away from export-oriented to domestically oriented growth. About half of the 11 percent growth in 2007 in China was generated by growth in the trade surplus and related investment, as explained below. A leveling off and progressive reduction in the trade surplus will likely slow the investment sector as well, which is deeply linked to export-oriented manufacturing industry and related infrastructure projects. Offsetting reduced export growth through increased private consumption, redirected investment, and higher government spending, however, faces a number of major obstacles.

How well India and China will cope with the various policy challenges and potential economic obstacles over the coming two to five years requires a more detailed look at the principal issues involved. For India, and then for China, eight major issues, which together will be decisive for the outcome, are examined, followed by a net assessment. The eight, labeled as to comparative advantage, are as follows:

- Education—*Advantage China.*
- Infrastructure—*Advantage China.*
- Agriculture/rural poor—*Mixed advantage.*
- Environment—*Advantage India.*
- Financial reform—*Advantage India.*
- Currency convertibility/exchange rate policy—*Advantage India.*
- A growing middle class—*Advantage India.*
- Democratization/rule of law/checks and balances on government power and corruption—*Advantage India.*

India Two to Five Years Ahead: A Sustained High-Growth Path

The discussion for India is straightforward. Each of the issues is examined in terms of what impact it will have on the ongoing path of balanced, high growth.

Education.—Education is one of the two principal potential obstacles to continued 8 to 10 percent growth in India over the next several years. The other is infrastructure. Progress is being made for both, but it is not yet adequate, and if the Indian economy stalls at less than 8 percent growth, the cause will almost certainly be in these areas. Education, including the need for reform as well as increased funding, was discussed in Chapter 2. Fundamental reform needs to begin at the primary school level, involving greater opportunities for expanded private schools and increased funding and restructured management for failed public school systems. The adoption of school vouchers could produce large positive results. University education needs to expand quantitatively for technology-oriented skills, where job growth is rapid, and qualitatively in terms of curriculum, faculty, and research facilities. This could require increased financing for public universities, including tuition payments by those who can afford it, and a more rapid expansion and accreditation for private colleges and technical schools. The danger signal to watch for in education is rapid wage increases for engineers and high technology-skilled workers. Wages have been rising rapidly for returnees from abroad and graduates from the top-tier schools, but the expansion of in-company training for the far more numerous graduates of lower-tier schools appears to be having a moderating effect on wage increases.

Infrastructure.—Increased infrastructure investment is critical for continued high growth. Electric power is the most important potential bottleneck, but a serious lack of transportation infrastructure—roads, railroads, airports, and seaports—is also impeding growth. Telecommunications infrastructure, in contrast, is not a problem in view of the rapid growth of private sector investment occurring in this sector. For example, Indian cellular operators plan to invest about $20 billion in 2007-2009 to provide mobile coverage for 80 percent of the population. Growth in electric gas and water supply was up from 5.3 percent in financial year 2006 to 7.4 percent in 2007 and to 8.3 percent in the first quarter of financial year 2008, but an adequate growth rate needs to be into double digits. Investment for transportation infrastructure appears to be rising, but reports are anecdotal, and aggregate figures would be valuable. One partial indicator of growth in infrastructure investment is the level of FDI in these sectors. During the first half of financial year 2007, compared with the first half of the previous year, FDI in the telecommunications sector was up by 950 percent to $405 million, in the electric power sector by 246 percent to $789 million, and in the transportation sector by 125 percent to $259 million.[83]

The danger signals for infrastructure are bottleneck delays and higher costs related to inadequate electric power and transportation. The limited information available as of late 2007 indicates that progress is under way, with investment growth likely accelerating for infrastructure, including electric power. As a positive example of the kind of anecdotal reporting that needs to be cumulatively assessed, the Chennai State-Owned Rural Electrification Corporation, in April 2007, extended $3.7 billion in credit at favorable rates to the local Tamil Nadu Electricity Board to expand power generation capacity and strengthen the transmission and distribution network over five years. Power distribution losses are a central problem for commercially viable power projects, averaging 33 percent nationally, while in Chennai the current 18 percent loss rate could be reduced to 7 to 8 percent or less from the new investment project.[84] Such an outcome would be real progress, with demonstration effect for other power projects in other states. In view of all of the foregoing, the assessment here is cautiously

[83] Ministry of Commerce and Industry, Press Release, November 20, 2006.
[84] See *Business Standard*, April 4, 2007.

optimistic that infrastructure will not become a crippling bottleneck impeding continued high growth.

Agriculture/rural poor.—This area includes the majority of the Indian population, where faster income growth and increased social services are needed for political as well as economic reasons. Although all segments of the population will benefit over time from 8 to 10 percent growth, as happened in China during the past 25 years, when the poverty rate declined dramatically and the literacy rate rose. But incomes in the modern, urban areas, including of the burgeoning middle class, grow relatively much faster, and steps to limit the consequent growing disparity in incomes are necessary as a matter of equity and, in India, for pro-reform political parties to win elections.

The Indian government places high priority on "inclusive" growth through financial transfers and investment incentives to stimulate faster growth and social improvements for the rural poor. The modalities involve comprehensive support for education, health care, and infrastructure, including potable water, and a more targeted sectoral strategy for agriculture. The government budget for financial year 2008 includes increases of 32 percent for education and 34 percent for health care, largely targeted on rural areas and low-income families, although the effectiveness of the additional spending will depend on improved program management. Infrastructure investment incentives include changes in regulatory constraints and targeted quotas for bank lending for agriculture and other rural projects. A less useful and often counterproductive component of public assistance is subsidies, such as low-cost or free electric power, which principally benefits higher income families and businesses while discouraging private investment to increase the power supply.

For agriculture, the share of the labor force on the farms has declined from 65 percent in 1990 to 55 percent in 2005, while the agricultural share of GDP has progressively declined to less than 20 percent, and these declines will continue, as they do in all countries during industrialization. The Indian agricultural strategy is responsive to this structural change, in part by providing improved education and infrastructure to enable the rural population to progressively transition out of agriculture, and in part by increasing farm productivity and shifting the composition of agricultural output to higher-value products. The development of more productive genetically modified crops was discussed earlier. The overriding immediate obstacle to higher farm incomes is the logistical failure to bring products to market. Forty percent of crops never make it to market because of poor roads, inadequate trucking, including an acute shortage of refrigerated trucks, and the lack of a fast-moving food distribution system. All of these elements of food-marketing infrastructure are receiving priority financial and other support to stimulate private investment. A decline in the share of crops lost in transit would bring immediate gains to farm income, while restraining inflationary food prices, and would create incentives for farmers to shift from basic food crops to higher-value, more labor-intensive food products, which are more vulnerable to loss in transit, and which are in growing demand by the rapidly growing middle class. The Indian food sector is projected to rise from $200 billion of sales in 2007 to $310 billion in 2015. The net result for farm income growth is hard to measure because of swings resulting from monsoon-related weather patterns. Agricultural growth was 3 to 4 percent per year during the 1980s but has recently been less than 2 percent. A substantial increase is clearly in order, and performance over the next couple of years needs careful monitoring.

Another major dimension of the strategy for alleviating rural poverty is to create more non-agricultural jobs, particularly for labor-intensive manufacturing and service industries. This relates to other areas of reform, particularly for labor laws that greatly inhibit such job creation for companies with more than 100 employees. The 2006 SEZ law, in particular, holds great promise for attracting such labor-intensive industry.

One other component of the outlook for agriculture and the rural poor is the potential role of private sector philanthropy by Indian business leaders and large corporations. The spirit of philanthropy and helping the poor is deeply rooted in Indian culture, as it is in other societies that are based on individual freedom and the market-oriented accumulation of wealth. For achievement-oriented "rugged individuals" in the United States, from Dale Carnegie and John Rockefeller earlier to Bill Gates and Warren Buffet today, and in India throughout the growing ranks of multimillionaire and billionaire private sector leaders, an important goal in life can become the desire to reinvest in society, for productive, life-enhancing purposes, much of the accumulated wealth. Among Indian companies, the Tata Group was again an early leader, when it established, in 1936, the Tata Charitable Trust, which receives 66 percent of corporate dividend distributions for charitable purposes to help the poor. Total philanthropic financing by the Tata Group in 2005, however, was only about $25 million, a relatively small amount given the current size of the company. Many other large Indian firms and wealthy individuals are beginning to provide support for projects to help the poor, and as revenues and profits soar, so, too, could philanthropic contributions. Moreover, large Indian companies in IT, communications, and manufacturing have the management and technical capabilities for mobilizing financial resources for quick and effective results.

The philanthropic dimension of support for the rural poor is not documented and does not yet appear to be substantial, but it could become so. If it were prominently recognized as producing significant results, it would, in addition to its direct assistance to the poor, help build national support for market-driven economic reform. There is also an opportunity for expanded collaboration between U.S. and Indian private sector philanthropy. A major collaborative initiative between the Gates/Buffet foundation and several of the largest Indian philanthropic leaders, on a matching-grant basis, could produce significant results and be well received by the low-income and impoverished majority of the Indian population.

Environment.—India and China face major environmental challenges and will have to devote increased resources for cleaner air, land, and water, which will have a growth-dampening effect. The subject has many important dimensions and no attempt is made here at an overall assessment. Rather, the basic point is made that the necessary expenditures and growth-dampening effects will be relatively much larger for China than for India. This is principally because the industrial sector is far larger and the course of pollution far more advanced and threatening in China. In addition, the more-open Indian society, with greater freedom for environmentalist groups and a culture deeply disposed to protect the natural habitat, provides a more favorable political setting for taking early and more effective steps to improve the environment, or at least to limit further environmental degradation.

Financial reform.—Prime Minister Singh has pledged early reform of the banking sector because "we cannot achieve our social and economic objectives unless there is reform of the insurance and banking system. Infrastructure requires long-term investment and our banking system is essentially short-term oriented."[85] State-owned banks still account for 75 percent of banking assets, although since the mid-1990s, private banks have grown in size and international reach, led by the Industrial Credit and Investment Corporation of India (ICICI) and the Housing Development Finance Corporation (HDFC). The banking system could expand private lending rapidly in response to reduced regulation, and banking services are of relatively high quality. One survey rated the Indian banking sector with respect to the regulatory, risk management, and technological systems, superior to those of China's and on a par with Japan's.[86] The pace of reform, however, has been slow, and much needs to be done to achieve the

[85] *Financial Times*, October 9, 2006.

[86] See "Challenges to International Banking," *FICCI Business Digest*, October 2006, pp. 10-14.

prime minister's objectives. The central challenge is to raise the very low "financial depth" of the economy, that is the ratio of financial assets to GDP. The majority of bank lending that does take place, moreover, goes to the government and to "priority" sectors, such as agriculture and household businesses, where there are limited creditworthy borrowers. Twenty-five percent of banking assets are required to be in government bonds, and another 35 percent is mandated for the priority sectors. International banks have very limited access to the Indian financial market. A comprehensive report on the Indian banking sector in 2006 by the McKinsey Global Institute concluded that "the government must loosen its grip on the financial system and allow financial institutions and intermediaries to respond to market signals." The report estimated that if this is done, Indian annual growth could be 3.4 percent higher.[87]

Banking reform to stimulate more bank lending for domestic investment will also lead to more Indian banking in international financial markets. India has the potential for becoming competitive in international financial services, following in the path of software and telecommunications services, and the three sectors are in the process of deepening their integration within India. Indians distinguish themselves in banking throughout the world, and the rise of private Indian banks during the past decade, together with a strong "credit culture," sets the stage for bringing together this expertise and experience to form an internationally oriented banking sector, with Mumbai as a major international financial center. This is the intent of the Indian government, and is contained in a high-level committee report in April 2007.[88] The report calls for the early convertibility of the rupee on capital account and greater competition and innovation throughout the banking sector in order to generate large exports of international financial services and to have Mumbai "develop the capacity to join London, New York, and Singapore as one of the premiere international financial centers of the world."

---

[87] See Diana Farrell, *et al.*, *Accelerating India's Growth Through Financial System Reform* (McKinsey & Company, May 2006). This 133 page report provides a comprehensive analysis of why India needs banking reform and how to achieve it.

[88] Ministry of Finance, "Brief Summary of the Report of the High Powered Expert Committee on Making Mumbai An International Financial Centre" (April 2, 2007). The 14-member committee included K. V. Kamath, CEO of ICICI, the largest Indian private bank; O. P. Bhatt, Chairman of State Bank of India, the largest public sector bank; Ravi Narain, Managing Director of the National Stock Exchange; and Usha Narayanan, Executive Director, Securities and Exchange Board of India.

The question is whether such reform will take place over the next several years, and the outlook is favorable for at least significant and possibly major change. Some changes are under way by stealth, for example through expanding the eligibility of bank lending that qualifies within the priority quota for agriculture to include roads and other infrastructure projects related to improved marketing of farm products. The Indian banking system, unlike the Chinese system, is favorably structured to adapt to the reforms, and there is broad support for moving forward. Major opposition comes from the unions of the overstaffed public sector banks, threatened by privatization and more intense competition from private banks, including foreign banks. The prospect of job growth throughout the banking sector, however, should enable all existing bank staff to remain employed. As for the big objective of making Mumbai an international banking sector, in addition to domestic banking reform, including far greater participation by foreign banks, the new international airport and connecting highway will be essential, and they are likely to be completed by about 2010. Other related financial issues are currency convertibility and exchange rate policy.

Currency convertibility and exchange rate policy.—The rupee was made convertible on current account in 1991, with the intent of making it convertible on capital account at a later date, but the action has been repeatedly postponed. The initial concern about capital account convertibility was that there could be a disruptive capital outflow and financial crisis as happened in 1991, but this is no longer a plausible argument. Trade is growing at about 30 percent per year in both directions, while the net inflow of long-term capital is substantial and growing, in response to the 9 percent GDP growth during the past several years. Moreover, the central bank has built up $300 billion of foreign exchange reserves, more than enough to counter a cyclical capital outflow and downward pressure on the rupee. The protective governess in Oscar Wilde's *The Importance of Being Earnest* cautioned her ward: "The chapter on the fall of the Rupee you may omit. It is somewhat too sensational." No longer. An early move to convertibility, through progressive raising and then elimination of annual ceilings on outward investment, would likely be absorbed by financial markets in an unsensational manner. The government's intent, as endorsed by the High Powered Expert Committee Report, calls for early convertibility on capital account, and a frequently cited date is 2009. Such a step would help pave the way for Mumbai to become an international financial center and for the rupee to become a major international currency.

A potentially more troubling issue concerns India's exchange rate policy. The current policy is a managed floating rate, and the issue is how it is managed. There are two basic options as the international financial system currently operates: the first is a market-oriented rate with central bank intervention limited to offsetting short-term developments in financial markets, while the second is protracted large-scale purchases of foreign exchange by the central bank to maintain an exchange rate below a market-oriented rate, which would make export industry more competitive. The second option is the currency manipulation issue discussed in detail in Chapter 8. Over the past several years, India, for the most part, has been going down the currency manipulation path, following the Chinese precedent which began in the late 1990s. India's foreign exchange holdings rose from $46 billion in December 2001 to $171 billion in December 2006, or by an average of $25 billion per year. By the end of 2007, foreign exchange holdings reached about $300 billion. During the same five-year period, the basic balance—that is current plus long-term capital account—which is the relevant measure of whether the exchange rate is market oriented, was in consistent surplus with large capital inflows outweighing current account deficits since 2003. In other words, in the absence of the central bank purchases, the rupee would have risen progressively in response to the net inflow of dollars from the basic balance surplus, while the central bank purchases have maintained the rupee below such a market-oriented rate. From 2001 to 2006, the rupee appreciated 9 percent against the dollar and depreciated 8 percent against the SDR, a currency basket rate. During 2007, however, with rising inflation, the rupee was permitted to float up by about 12 percent.

The fact is that for high growth, rapidly industrializing economies that are open to trade and foreign investment, as are China and India, a market-oriented exchange rate will rise over time in response not only to the net inflow of FDI and other long-term investment, but also to the consequent very high productivity growth in manufacturing and service industry exports. This progressive appreciation of the currency—or terms of trade—can bring great benefits from lower import prices, restraint on inflation, and increased purchasing power for consumption and travel abroad. The immediate disadvantage is the price squeeze on export- and import-competing industries, as the exchange rate works to reduce the basic balance surplus. The currency manipulation option, in contrast, provides a competitive advantage for exports, while not only losing the above-listed benefits, but in having the central bank divert large amounts of national savings to invest abroad in U.S. Treasurys and other financial assets, rather than at home. For India, investing at home would mean providing more inclusive benefits for the majority of the population who are very poor.

Indian exchange rate policy, while following the Chinese path, is still at the early stages of the currency manipulation option, but it is a slippery slope that could soon lead to excessively export-oriented growth and the kind of macroeconomic distortions shown in Figures 5-1 to 5-4 for China. A major constraint on a more market-oriented exchange rate policy for India, however, as for East Asian economies, is the far more undervalued Chinese currency which, as recounted earlier, has led to a large and growing trade surplus with India. At this point, India is generally pursuing a heavily managed floating rate policy to keep the exchange rate below a market-oriented level, with limited public discussion of the Indian interests involved, although, as noted above, the rupee was permitted to rise during 2007. The international interest, including that of the United States, is left for Chapter 8. It is sufficient to conclude here with a rephrasing of the advice from Oscar Wilde's governess: "The chapter on the rise of the Rupee you should not omit. It could become highly sensational."

The rising middle class.—This is a central and critical dimension for the transformation of India and China from overwhelmingly poor developing countries to advanced technology superstates. The end result will be that the majority—ultimately the large majority—of the population becomes educated, relatively affluent, and highly productive, while in the process becoming the dominant force within the economic and political power structure of each country. The estimated size of the middle class depends on definition and measurement techniques, but rough orders of magnitude are sufficient for understanding the forces in play. Estimates of the Indian middle class range from 100 to 200 million in 2007, with a corresponding Chinese middle class of 150 to 250 million. The estimated annual growth of the middle class is 20 to 30 million in each country. This means an Indian middle class currently at about 10 to 15 percent of the population, rising to 20 to 30 percent over the coming ten years. These absolute numbers, moreover, greatly understate the relative economic and political power wielded by the more highly educated and economically powerful middle class.

For India, the rapid growth of the middle class is an unqualified advantage for continuing the economic reform strategy and strengthening the roots of democratic government and the rule of law. The self-interest of the middle class lies predominantly in these directions, while the demonstration effect of tens of millions of Indians rising each year from grinding poverty to the joys of modern housing, greatly enhanced personal consumption, and the opportunity to provide a good education for their children, is increasingly evident throughout the country. As the new middle class extends its location to the outlying cities, this demonstration effect will become even more direct. And most profound in historical context, the expansion of the middle class through performance-based education and private sector job creation is reducing the caste and class barriers that have been the great stain on Indian culture. Class conflict will continue in India, most severely in the poorest regions, but the steady expansion of middle-class incomes and values will progressively become the preferred, nonviolent route for bridging the caste divide.

One other important shift in the balance of power in India resulting from the private sector driven expansion of the middle class is the relative decline within the power structure of the government bureaucracy—the license Raj—which is a pervasive obstacle to deregulation and economic freedom. Until only 15 years ago, the choice for able college graduates was still basically between the relatively few lifetime jobs in government service and emigration to find professional work abroad. Now the large majority of graduates, and particularly the most highly motivated, enter the private sector, where their interests tend to clash with those of the license Raj. The balance of power is shifting, and the results are increasingly evident of a bureaucracy more on the defensive and threatened by absolute decline through privatization and the job-displacing application of new information technologies.

Democracy/rule of law/checks and balances on government power and corruption.—A strengthening of democratic process follows from almost all of the foregoing. A growing middle class brings greater political stability and support for economic reform. The larger relative power of the private sector constrains the abuse of power by the government. A more educated population becomes more engaged in public policy toward better governance. The steady rise of incomes for the poor majority, even though too slowly, still raises hope for peaceful democratic change rather than antidemocratic violence.

The central role of technological advance in general, and of applied information technologies in particular, is also supportive of the rule of law and the strengthening of checks and balances on government power and corruption. In India, it has enabled effective implementation of the 2005 Right to Information Act, similar to the U.S. Freedom of Information Act, which is providing greater transparency for government process. Nongovernmental organizations are actively using the law to uncover government abuse. Criminal trial proceedings, including unpunished crimes against lower-caste victims, are being exposed by the media, with highly publicized convictions that act as a deterrent to such crimes. Electronic access to full information on competitive bidding procedures for public sector investment projects is acting as a constraint on long-standing corruption in this area. Freedom of the press in India, now extended to multi-channel television and the Internet, is open to all and is flourishing, to positive democratic effect.

The ultimate political test will be whether the rates of crime and government corruption go down, and whether other basic measures of good governance go up. Definitive data are not yet available, but the directions of change are clearly positive. The fundamental analytic point is that market-oriented, high economic growth, including new information technology applications, results in a number of mutually reinforcing consequences in support of strengthened democratic governance under the rule of law.

* * *

These are the various issues in play as the economic reform and modernization process moves forward in India. There are potential obstacles to continued high growth. Much larger investment in infrastructure and expanded quality education are the most daunting immediate challenges. More rapid benefits for the poor majority and banking reform are also essential. The many positive developments, however, have been mutually reinforcing and provide momentum to meet these challenges. The net assessment here is thus that this reform momentum is highly likely to continue over the coming two to five years so as to maintain the 8 to 10 percent growth path.

China Two to Five Years Ahead: A Hard Landing Adjustment

The two to five year projection for China requires a fairly specific scenario about structural adjustment away from export-oriented to more domestically oriented growth. This is the stated objective of the Chinese government, with strong support and growing pressures from trading partners, particularly the United States and the EU, and from within the IMF system. The resulting scenario adopted here is that the surging Chinese trade surplus, up by almost 50 percent during 2007, together with a current account surplus of 12 percent of GDP, will level off and begin to decline over the coming two to five years. This, in turn, as described below, will require major actions by the Chinese government, most importantly a large revaluation of the yuan. The consequent adjustment, in terms of increased domestic demand and a smaller trade surplus, will be extremely difficult, with transitional adverse impact on some sectors and regions. The term "hard landing" is thus used to characterize this adjustment away from a severely unbalanced economy, trending toward even further imbalance, as demonstrated in Figures 5-1 through 5-4.

Problems in reducing the current account surplus and the consequent $500 billion per year of central bank purchases, and in stimulating an offsetting increase in domestic growth, are elaborated throughout the eight issues discussed below. Two basic points are made at the outset, however, which indicate the daunting scope of required restructuring of growth. Together, these two points lead to a projected scenario of reduced growth, perhaps to 5 to 7 percent, at least for a couple of years, together with substantial adverse impact on some manufacturing industries.

The first point is that about half of recent growth has been generated by the surging trade surplus and by investment in export industry and related infrastructure, and this growth-generating effect will have to shift into reverse. In 2007, the trade surplus increased by $85 billion, or about 3 percent of GDP, while related investment growth probably added another 2 to 3 percent. Thus, about 5 to 6 percent of the overall 11 percent growth was caused by the increase in the trade surplus. Moreover, the projected adjustment to a decline in the trade surplus of perhaps $60 billion, or 2 percent of GDP, per year, would have a further dampening effect on growth and lead to an overall net decline in trade-related growth of 7 to 8 percent. Some of this decline in growth would presumably be offset by increased domestic growth, but in view of the very low, 36 percent personal consumption share of GDP in 2007, and other factors elaborated below, the additional domestic growth would probably be relatively small, perhaps, at best, adding 3 to 4 to overall growth during the first couple of years. This would bring the net growth rate down from recent 9 to 11 percent levels to the 5 to 7 percent range.

The second point relates to the adverse impact the projected adjustment would have on manufacturing industry, starting from the extremely large trade surplus in manufactures of $450 billion in 2007. To reduce this surplus, imports would have to grow at a much faster rate than exports. For a significant decline in the surplus, imports would have to grow on the order of 25 percent, compared with 15 percent for exports, the reverse of recent trade performance.[89] Such reversal would mean not only a sharp drop in export growth, 95 percent of which is in manufactures, but also a significant increase in the import share of a slower-growth domestic market for manufactures. Moreover, a decline in export growth would have a corresponding negative impact on imports of components for reexported products, the export platform effect, which would mean an even larger percentage

[89] In 2005, exports grew 24 percent and imports 13 percent; in 2006, exports grew 24 percent and imports 16 percent; and in 2007, exports grew 26 percent and imports 21 percent.

increase in the import share of the domestic market for manufactures of final products. But aside from some higher tech capital goods and commercial aircraft, it is hard to see where imports could gain domestic market share without causing significant adverse impact on domestic industry, already facing overcapacity in a number of sectors, and faced with much slower export growth. For example, what would be the impact on the steel, automotive, semiconductor, consumer electronics, pharmaceutical, apparel, and footwear industries from an increase in the import share of the domestic Chinese market, at the same time that export growth was sharply reduced? In this context, a hard landing is certainly an appropriate characterization of what likely lies ahead for the manufacturing sector if the trade surplus is to begin to decline.

The response to this prospective dilemma for Chinese manufacturing could be that China simply will not permit reductions of its trade surplus and current levels of purchases of foreign exchange by the central bank. This issue is addressed in Chapter 8, and the discussion here of the eight issues is simply an assessment of the course ahead predicated on the assumption that the official objective of restructuring away from excessively export-oriented growth will, in fact, occur over the coming two to five years.

Education.—This is one area where China, in contrast to India, has the resources in place to support the transition to more balanced growth, although changes in direction will be required, and it is not clear how quickly and effectively the Chinese government will respond. There is already a surplus of engineers graduating from universities, and the shift away from export-oriented growth in manufactures will mean an even slower growth in engineering jobs. Job creation for domestic growth will be more in the direction of health care workers and other service industries, but the restructuring of Chinese universities away from engineering and in these new directions will not come easily.

Infrastructure.—Again, China is devoting more than ample financing to infrastructure, but change in direction will be required. The financing will need to shift away from the heavy concentration in highly industrialized, export-oriented port cities to provincial urban and rural needs, and to environmental objectives, including cleaner energy production. The incentives for such shifts in infrastructure investment relate largely to the responsiveness of the troubled banking system, discussed below.

Agriculture/rural poor.—The Chinese challenge in this area is largely different from that of India, but daunting nonetheless. The rural labor force of 485 million, in 2005, was 62 percent of the total labor force, of which 340 million, or 44 percent, was in agriculture. The poverty level has been greatly reduced in China, but the rural population, as in India, has not fully shared in the recent 9 to 11 percent annual growth. The agricultural sector accounted for 13 percent of GDP in 2005 and grew at less than 5 percent per year during 2003-2005, or less than half the national average growth. The government commits substantial resources for agricultural research, including for more productive genetically modified crops, described earlier. Broader challenges for growth in agricultural production are the relative scarcity of arable land and the shortage of water resources, the latter including bad distribution, poor efficiency, water pollution, and contradictions in the way water is allocated and managed.[90] Greater investment and technical resources in these areas would not only address the needs for increased agricultural production but would also contribute to higher overall domestic growth.

The more immediate and serious problem for the rural and agricultural populations in China is land ownership and the appropriation of farmlands for industrial and urban development. It is the same difficult problem that is being faced in India, but the Chinese response is fundamentally different, reflecting the difference in political systems. In India, as recounted earlier, violent street demonstrations by farmers protesting the new SEZ law with respect to the purchase of farmland led to massive media coverage and threatened lawsuits, and the federal and state governments responded by revising procedures and the law to ensure that farmers receive fair compensation for their land plus other benefits. In China, in contrast, in the absence of a free press and an independent court system, a confrontational process plays out in the streets, year after year, including the destruction of property and frequent police violence. Occasionally local party officials are fired for permitting the protests, under the policy of "responsibility systems," which holds local officials responsible for allowing demonstrations to take place, resulting in even harsher repression to dissuade potential protesters.[91] There were more than one million illegal land seizures between 1998 and 2005, and the number of public protests rose from 74,000 in 2004 to 87,000 in 2005. The overall process of appropriating farmland without legal recourse is viewed by the rural population as highly corrupt and unjust to farmers, and this, in turn, limits the prospect for community-based initiatives to increase domestic growth. It is also an important example of the conflict between market-oriented economic growth and authoritarian political rule, particularly in the more distant regional cities and the rural population.

---

[90] See OECD, *Environment, Water Resources and Agricultural Policies: Lessons from China and OECD Countries* (OECD, 2007).

[91] See Carl Minzner, "Social Instability in China: Causes, Consequences, and Implications," *The China Balance Sheet in 2007 and Beyond*, C. Fred Bergsten, *et al.*, eds (Peterson Institute for International Economics and CSIS, May 2007), pp. 55-78.

Environment.—China faces enormous environmental challenges in terms of land, air, and water pollution. In 2007, China likely passed the United States to become number one in greenhouse gas emissions, largely because 70 percent of Chinese electricity is generated by exceptionally dirty, coal-fired generators. Poorly regulated industrial plants pose an environmental threat to waterways, and increased air pollution in Beijing during the 2008 Olympics is a concern. Again, as for India, no attempt is made here to assess the economic costs of the necessary response to these growing environmental hazards, except to state that in relative terms they are far larger for China than for India. In addition, the Chinese political setting, plagued by corruption and secrecy that delays the reporting of environmental mishaps, is less favorable for cost-effective responses to environmental degradation.

Financial reform.—The need for radical reform of the Chinese financial system—banking, insurance, and capital markets—is probably the most important economic obstacle to restructuring growth away from manufactured exports to more broadly-based domestic growth. Export industries have relatively small needs for the domestic banking system. Investment can be financed internally through export earnings or through borrowing abroad. In contrast, companies serving the domestic market—smaller manufacturing companies, service industries, and agribusiness—are far more dependent on Chinese financial services, and these services are woefully inadequate. The problem is recognized, and comprehensive financial reforms have been promulgated, but implementation is very slow, for political as well as economic reasons.

In the banking sector, the four large state-owned commercial banks, in 2005, accounted for 54 percent of total bank assets, and another 15 shareholding and policy banks held an additional 24 percent. These greatly overstaffed public sector banks misallocate lending at great cost to the domestic economy, with large numbers of nonperforming loans and an even larger number of underperforming loans whose marginal returns are possible only because of the large spread in government-fixed interest rates between lenders and borrowers. Almost three-quarters of bank lending goes to state-owned or collective enterprises, which produce one-quarter of national output, whereas private companies produce well over half of total output while receiving only one-quarter of bank loans. An overarching institutional problem of the state-owned banks is that bank staff lack a "credit culture" and have limited technical capability for risk assessment and loan management.

Other financial services in need of far-reaching reform are insurance and capital markets. A major reason the Chinese savings rate is so high and personal consumption so low is that there are few available health care or retirement annuity programs available, in the public or private sectors. People self-insure without spreading the risk, which means much higher average savings rates in the absence of a functioning insurance industry. As for capital markets, they are exceedingly small, and almost all credit comes from commercial banks, which adds to their noncompetitive behavior. In 2004, equity capitalization as a percent of GDP was 1 percent in China, compared with 56 percent in India, 79 percent in Japan, and 139 percent in the United States. Corporate bond markets in China are even smaller.

A 2006 study of the Chinese financial system by the McKinsey Global Institute concluded that implementation of a range of financial reforms would increase investment efficiency and raise GDP by 13 percent.[92]

The prospect is for continued slow progress in implementation of financial reforms, and therefore financial services will remain a major obstacle to raising domestically oriented growth over the coming two to five years. The problem is partly technical, the lack of professional banking expertise and a credit culture, but it is also highly political. The Chinese banking system and its massive financial relationship with state-owned enterprises involve deeply embedded and widespread corruption and nepotism, on both sides of the relationship. Bank reform leading to greater efficiency and competitive pricing would jeopardize the solvency of both the state-owned banks and enterprises. Another major problem for the banking system is the need to absorb, or "sterilize," the central bank's huge purchases of foreign exchange so as to restrain inflation. Finally and most importantly, continued Communist Party control of the economy and the country depends critically on continued control of financial resources.

Exchange rate policy and currency convertibility.—The order of the terms is reversed for China because exchange rate policy is far more important while currency convertibility remains a distant prospect. Exchange rate policy, through appreciation of the yuan, will be the most powerful policy instrument for reining in the runaway Chinese trade surplus. It will also have a major indirect effect on shifting resources toward domestically oriented growth, as relative profits for exports decline compared with domestic sales. Chinese policy is to move to a market-based exchange rate, but so far nothing significant

---

[92] See Diana Farrell, *et al.*, *Putting China's Capital to Work: The Value of Financial System Reform* (McKinsey & Company, May 2006). This is the companion piece to the study of the Indian financial system cited earlier, and the two together highlight important differences between the two systems. This estimate, moreover, was based on the export-oriented growth path at the time, and the relative GDP gains from financial reform with more balanced, domestically oriented growth would be substantially larger in 2007.

has happened, and pressures are building from trading partners, especially the United States and Europeans, for a substantial revaluation of the yuan. This policy confrontation is addressed at length in Chapter 8. The central question underlying the policy debate is the degree to which the yuan is undervalued, an elusive question which has been confused by frequent groundless and misleading statements. An orderly presentation of the relevant information for making a best estimate, or order of magnitude, of the undervaluation of the yuan is therefore vital for a more productive policy discussion, and is offered in the discussion that follows.

The short answer to the question is that the yuan is greatly undervalued, but there is no way to measure the undervaluation with any precision. The conclusion drawn from what follows is that, in 2007, the yuan was undervalued by at least 50 percent, and probably much more, compared with a market-oriented rate, but this should be considered an order of magnitude to emphasize that the adjustment will have to be far larger than is implied in official discussions about a modest revaluation of 10 percent or 20 percent, which is unlikely to halt the continued rise in the Chinese current account surplus.

A point of departure for the assessment is the experience of Japan in the 1960s and the 1970s, and of the United States in the 1980s. The Japanese yen appreciated 300 percent over 20 years, from 360 yen to a high of 90 yen to the dollar, as large investment in export-oriented manufactures, incorporating more advanced, cost-cutting technologies, led to large increases in productivity and highly profitable export growth in manufacturing industry. China has been following a similar path since about 1995, but at an even faster pace than Japan because of the central role of foreign investment which was absent in Japan. Thus, based on market forces, the Chinese yuan might rise much as the yen did, and if not by 300 percent, at least by 100 to 200 percent.

The dollar declined by 40 percent in the mid-1980s, a decline that, in turn, brought about a decline in the U.S. current account deficit of about 3 percent of GDP after three years. A considerable amount of empirical testing of this trade adjustment experience was undertaken, leading to a widely accepted assessment of a "10:1 benchmark," that is, it takes a 10 percent decline in the dollar exchange rate to bring about a 1 percent of GDP decline in the current account deficit. Applying the 10:1 benchmark for China in 2007, with a projected current account surplus of 12 percent of GDP, or three times larger than the U.S. deficit in the 1980s, might appear to indicate a 120 percent revaluation of the yuan to bring the current account into balance, but the circumstances in China are different in at least a couple of ways. There is first the asymmetry between the percentage declines for a depreciating currency and the much larger percentage increases for the appreciating currency. For example, when the dollar declined by 40 percent in the mid-1980s, other currencies appreciated by about 50 to 70 percent.[93] Thus, whatever the benchmark, the percentage appreciation of the yuan would be much larger than the corresponding depreciation of the dollar. Another factor is the flexibility of the economy in question to adjust external and internal accounts. IMF analysis of this question concludes that countries with less-flexible economies may need larger movements in their exchange rates to adjust trade imbalances. Certainly China, at this juncture, faces great inflexibility in its capacity to adjust, as described in the other issues here, particularly banking reform. In any event, however the American experience in benchmarking relates to China, a very large percentage adjustment, again in three figures, is indicated.

A series of direct estimates of the undervaluation of the yuan were made in 2003 and 2004, using different measuring techniques, with results generally ranging from 25 to 70 percent. Financial guru Henry Kaufman estimated that China would have to revalue its currency upward by 50 to 70 percent in order to realign its economic and financial relationships.[94] Jeffrey Frankel used differences in purchasing power parity (PPP) to estimate that in 2005 the yuan was undervalued by at least 45 percent in logarithmic terms and 36 percent in absolute terms.[95] Morris Goldstein estimated an undervaluation in 2004 "in the upper half of the 15 to 30 percent range," but this range was understated because it was based on the first months of trade in 2004, which projected a decline in the Chinese trade surplus, whereas the surplus actually rose by over 30 percent for the entire year.[96] The author estimated an undervaluation in the order of 40 percent, related to the experience of the United States and other trading nations in the 1980s and the degree of imbalance measured in terms of basic balance rather than current account balance.[97] All of these estimates from 2003

---

[93] As a more precise example of this asymmetry, if the dollar declines from two pesos to one peso to the dollar, it constitutes a 50 percent decline for the dollar, but for the rising peso—up from 50 cents to one dollar to the peso—it is a 100 percent appreciation.

[94] *Financial Times*, December 9, 2004.

[95] Jeffrey Frankel, "On the Renminbi: The Choice Between Adjustment Under a Fixed Exchange Rate and Adjustment Under a Flexible Rate" (NBER Working Paper No. 11274, April 2005).

[96] Morris Goldstein, "China and the Renminbi Exchange Rate," C. Fred Bergsten and John Williamson, eds., *Dollar Adjustment: How Far? Against What?* (Institute for International Economics, 2004).

[97] Ernest H. Preeg, "Exchange Rate Manipulation to Gain an Unfair Competitive Advantage: The Case Against Japan and China," C. Fred Bergsten and John Williamson, eds., *Dollar Over Valuation and the World Economy* (Institute for International Economics, 2003), pp. 267-284. This was the earliest of the estimates, and the 40 percent figure has since taken on a life of its own.

and 2004, however, would have to be approximately tripled, to a range of 75 to 210 percent, in view of the tripling of the Chinese current account surplus from 4 percent of GDP in 2004 to 12 percent in 2007.

These are the various lines of analysis that bear on how far the yuan is undervalued. They all point to a very large undervaluation, which should come as no surprise in view of the large, skyrocketing Chinese trade surplus. Nothing like this has happened before, and the degree of adjustment that will be needed to reach a market-based exchange rate is likewise without precedent. In this context, the assessment here that the yuan is undervalued by at least 50 percent, and probably much more, is, if anything, an understatement.

Finally, a brief comment on currency convertibility. Convertibility of the yuan on capital account is not a viable option for China as long as the exchange rate is so greatly undervalued. Nonconvertibility greatly limits the use of the yuan as an international currency and therefore also limits the rise of Shanghai, compared with Mumbai, to become an international financial center.

The rising middle class.—This is a central and critical dimension in the transformation of China to an advanced technology superstate, but unlike in India, where the rapidly growing middle class is an unqualified plus in economic and political terms, it is a mixed blessing for the Chinese communist government. On the positive side, an educated, relatively affluent middle class of 150 million to 250 million in 2007, projected to double in 10 to 15 years, provides the human resources for the continued high growth of a modern, industrialized society. It is also the source of higher growth in personal consumption, which is key to the transition from export-led to domestically led growth.

The potential negative consequence for the Chinese government, however, is the threat that the rise of the middle class to become the dominant center of the economic and, ultimately, political power structures of the country, poses for the Communist Party to continue one-party authoritarian rule. This threat, moreover, is both technical and political. The technical challenge is one of effective management. It becomes harder and harder to direct and control a rapidly evolving, market-oriented economy, driven by hundreds of millions of middle class participants pursuing their economic self-interest, through an all-powerful Politburo Standing Committee of nine at the center and then out through government ministries, state governments, and local Communist Party cadres.

The political challenge is how to respond to an educated and economically ever more independent middle class that wants to participate in the way government powers are exercised—in other words, that wants democratization. The globalization of the Chinese economy, in terms of foreign companies widely dispersed throughout the country, the growing travel abroad by middle-class Chinese, and the 170,000 Chinese who have returned from living in the democratic diaspora, all reinforce the political interest of the Chinese middle class in democratic change.

Democracy/rule of law/checks and balances on government power and corruption.—This is the fundamental political issue for determining the course ahead for China, and there will likely be some change over the coming two to five years. It is also the most difficult area to predict, with so many forces in play, and a detailed assessment is beyond the scope of this study. A few basic observations, however, are offered, particularly in connection with the economic restructuring challenge described in the previous issues.

Premier Wen Jiabao, in his March 2007 annual press conference, stated, "Democracy, the rule of law, freedom, equality and fraternity are not something peculiar to capitalism. These are also the common values that we as human beings all pursue."[98] The Chinese government has thus set democratization as its goal, but has no specific schedule or program to achieve it, and since 2005 authoritarian controls, on balance, have been tightened. An earlier program for elections at the local level has been restricted so that designated party candidates almost always win, while media reporting and Internet communication have been curtailed and the courts remain firmly under government control. Increased repression of the rural population through the "responsibility system" was described earlier.

The principal observation made here is that the political difficulties of managing an ever more market-oriented, high-tech economy will increase further as the government is forced to confront a restructuring of growth away from exports to the domestic market. The adverse impact of such restructuring on particular industries or regions will elicit strong criticism and public protest. An overall slowdown in the economy and a broadly based squeeze on the manufacturing sector will intensify the criticism, especially from the powerful leadership of export industry, which has been reaping huge profits from the undervalued currency. History demonstrates that political change is more likely in times of trouble than of prosperity, and troubled times are likely ahead.

The pervasive problem of corruption and nepotism will also grow more threatening as the economy diversifies and decentralizes, whereby growth is restructured away from a few export-oriented industrial centers. Lord Acton's sage warning that power corrupts and absolute power corrupts absolutely has never before applied to a multitrillion dollar economy, and a difficult period of structural change, with uneven impact, will increase the incentives for corruption and sharpen the

---

[98] *Washington Times*, March 17, 2007.

public's discontent with it.[99] The only feasible antidote to corruption and nepotism is independent checks and balances on government abuse, which goes to the core of the democratization process.

A final comment about current Chinese authoritarian rule concerns the character of decision making, which currently falls far short of what is needed to respond to the challenge of structural imbalance and massive misallocation of investment through the state-controlled banking system. The economic performance of President Hu and his Politburo has been largely a denial of the problems at hand and the habit of highly risk-averse decision making, leading to a lowest common denominator response to problems. As summarized in a widely cited assessment, "In contrast to his predecessors—Mao, Deng, and Jiang—there is no 'thought,' 'theory,' or 'important thinking' officially ascribed to Hu Jintao—yet."[100] This absence of thought is in sharp contrast to the coherent, action-oriented decisions being taken by the Indian economic reform team, despite severe constraints within an unwieldy democratic coalition government.

The great ideological uncertainty facing President Hu and the Communist Party is whether China will in fact change political course toward democratic government under the rule of law or will continue unyielding authoritarian control. The coming two to five years of difficult economic restructuring will only intensify the forces for democratic political change, but how the government will respond cannot be predicted.

* * *

These are the various issues whose cumulative impact leads to the assessment that China will undergo a difficult adjustment to more balanced growth over the coming two to five years, and is highly likely to experience at least a couple of years of slower growth, perhaps down to 5 to 7 percent. The uncertainties as to the precise course of events are far greater than for the Indian assessment, but the basic conclusion for China is fairly certain. There will be a relatively hard landing adjustment to more balanced growth. The requisite structural adjustment faces a number of daunting obstacles, and at least some

---

[99] See Minxin Pei, "Corruption Threatens China's Future" (Carnegie Policy Brief 55, October 2007). Pei estimates that the direct costs of corruption were $86 billion in 2003, or 3 percent of GDP. With a PPP adjustment, the purchasing power of bribe takers rises to over $200 billion, far beyond precedent for any one nation in one year.

[100] C. Fred Bergsten, *et al.*, *China: The Balance Sheet* (Institute for International Economics and CSIS, 2006), p. 60.

sectors and regions will suffer substantial adverse impact. A critical factor is the ability of the Chinese government to facilitate the transition through coordinated, forceful action, but based on performance over the past four years, such action is unlikely to be forthcoming.

One final important development over the coming two to five years for the relative growth paths of China and India is the exchange rate relationship. As explained earlier, newly industrialized, export-oriented economies achieve high GDP growth through very high productivity growth, particularly in export-oriented manufactures, which should lead to large upward movements in the exchange rate, as in the 300 percent rise of the Japanese yen. A large upward movement of both the Chinese yuan and the Indian rupee is virtually inevitable,[101] but because the Chinese policy in recent years has been to maintain a greatly undervalued yuan through massive central bank purchases of foreign exchange, a much larger and earlier catch-up revaluation of the yuan will occur, almost certainly beginning during the coming two to five years. The largest relative beneficiary of this revaluation could well be India. India, increasingly, is competing with China across the board for FDI and export markets, in manufacturing and business service industries, including for attracting R&D programs by MNCs. A major revaluation of the yuan vis-à-vis the rupee would make India a relatively more attractive place to invest. This would even include an outward flow of Chinese FDI to India, especially for labor-intensive industry, which will initially be the hardest hit from a yuan revaluation.

The Longer Term Outlook

Longer term projections of 10 to 20 years are subject to far greater uncertainty than the preceding two to five years assessments. Nevertheless, the broad lines of longer term development, and the most important issues of potential change, are fairly clear. Two basic questions are addressed here. The first question is whether China and India will rise to become advanced technology superstates, approaching parity with the United States. The definition of an emerging advanced technology superstate, introduced in Chapter 1, is a nation which almost inevitably will achieve giant economic, technological, and financial status, will very likely become financially and politically powerful in international affairs, and will eventually strive to become a military superpower as well. The assessment here is that China is

[101] An alternative route would be a sustained high rate of inflation, but this appears unlikely for both China and India.

already close to being a fully engaged advanced technology superstate, and will almost certainly rise toward parity with the United States over the coming 20 years. India, in contrast, while well launched toward such superstate status, is still at least 10 years away from achieving it in economic terms, and even further away as a military superpower. Over time, however, and certainly over 20 years, India should progressively rise to superstate status, and at a fairly rapid pace, as long as the Indian economy remains open to trade and international investment. This outcome is not projected in detail. The economic forces in play have already been discussed at considerable length, and military modernization for both countries is addressed in the following chapter.

The answer to the first basic question in systemic terms, therefore, is that, over the coming 20 years, the world political and economic orders will increasingly center on three advanced technology superstates—the United States, China, and India—with Japan an ever more distant fourth place. The EU, although a global economic superpower, comparable in size to the United States and China, is not a political state, and currently lacks the foreign policy and military dimensions of superstate status. The far-reaching implications of this emerging relationship among four "regional economic hegemons" are examined in detail in Chapters 7 and 11.

The second basic question is how long China and India are likely to maintain the very high, 8 to 11 percent annual growth. The assessment here is that at some point over the coming 20 years, as industrial modernization matures and broadens to include most of the economy, sustainable growth will moderate, perhaps to 4 to 6 percent. This is what happened earlier to Japan, South Korea, and Taiwan. The dynamic model underlying this assessment is that, assuming 1 percent annual population growth, 8 to 11 percent GDP growth means 7 to 10 percent annual growth in productivity, which requires ever larger amounts of new, productivity-enhancing investment. And as the majority of the economy becomes modernized through investment with applied state-of-the-art technologies, a national average increase in productivity of 7 to 10 percent can no longer be sustained.

The experience of Japan, South Korea, and Taiwan involved a 25 year cycle, with high levels of investment and economic growth peaking out after 15 to 20 years, followed by a transition to lower levels of investment and growth.[102] China began its advanced technology industrial takeoff in the early to mid-1990s, which indicates that it is now approaching the 15-year mark. Chinese industrial modernization has also been moving at a faster pace because of the

[102] See Lardy, *op. cit.*, 2007, pp. 2-3.

major role played by FDI, which was not part of the earlier experience of the three noted economies. India, in contrast, only began a broadly based industrial modernization, including the manufacturing sector, during the past two to four years.

In this dynamic growth context, the assessment here is that China will reach the point of more moderate growth far sooner than India, and probably within 5 to 10 years. Half of the Chinese economy, as measured by GDP, is already in the industrial sector, while structural adjustment problems and other constraints on growth will have a cumulative additional growth moderating impact.

India, in contrast, is at a much earlier stage of industrial takeoff, with manufacturing and related service industries relatively small and ripe for extended rapid growth, for both the export and the domestic markets. India has been growing at 9 percent per year for four years, with investment as a share of GDP rising to a robust 34 percent. Very broad investment opportunities for infrastructure, manufacturing, and new service industries should thus be able to sustain or even increase the 9 percent growth path for another 10 or perhaps 20 years.

India also has a faster-growing population than China and will probably surpass China in population within 20 years, but this is a mixed blessing. More young Indians will enter the labor market, but this means greater demand for schools and job creation. The assessment here is that the supply of labor will be more than ample over the coming 20 years in both countries as hundreds of millions of people make the transition from rural poverty to urban middle-class life. Moreover, in the process, more numerous and healthier senior citizens can work a few years longer.[103]

As for broader relationships, China and India, whatever their growth trajectories, will continue their advance toward joining the United States as advanced technology superstates and thus becoming two of the three corners of a new Asia-Pacific triangle, which has wide-ranging implications for both the world political and economic orders. The second half of this study addresses these implications and offers a comprehensive U.S. policy response, both for immediate and longer term U.S. interests and objectives. First, however, the geopolitical and geostrategic dimensions of the recent rise of China to advanced technology superstate status, and the incipient similar takeoff by India, need to be addressed, including how they relate to others in Asia, and to the United States.

---

[103] As role models, both Prime Minister Singh and the author are leading reasonably productive lives into their eighth decade.

CHAPTER 6

The Geopolitical and Geostrategic Dimensions

The analysis of China and India as rising advanced technology superstates, up to this point, has focused on the economic and commercial dimensions which are the first part of the superstate definition, "the achievement of giant economic, technological, and financial status." The second and third parts of the definition, however, are also of far-reaching importance: "becoming financially and politically powerful in international affairs," and "striving to become military superpowers." They are the subject of this chapter, labeled broadly as the geopolitical and geostrategic dimensions. The geopolitical dimension is discussed here in terms of regional relationships within Asia, leaving the global relationships for later chapters. In particular, the rise of China to become the economic hegemon in East Asia, and India's growing role of bringing greater balance and broader geographic scope to regional relationships within Asia, are examined. As for the geostrategic dimension, the rise of China to become the dominant Asian military power is discussed first, followed by a less definitive commentary on India's military capability and aspirations.

China as Economic Hegemon in East Asia

In a little more than a decade, from 1995 to 2006, China rose to displace Japan as the predominant economic power in East Asia, as shown in Table 6-1 for trade. In 1995, Japan accounted for 37 percent of the exports and 30 percent of the imports of the East Asian total, while the Chinese shares were about half as large, at 15 percent and 17 percent, respectively. By 2006, the roles had reversed, with China accounting for 33 percent of exports and 31 percent of imports, compared with 22 percent and 23 percent for Japan, and China's lead position widens each year, as Chinese trade grows at double the Japanese rate.

China's rise to primacy in regional trade is particularly important for the Japanese trade relationship with the United States, its number one trading partner for over 50 years. As shown in Table 6-2, in 1995 28 percent of Japanese exports and 23 percent of imports were accounted for by the United States, while the Chinese shares were less than half as large, at 8 percent and 11 percent, respectively. By 2006, China had surpassed the United States in terms of Japanese imports, at 21 percent of the total compared with 12 percent for the United States, and although Japanese exports to the United States still exceeded those to China, the U.S. share of total exports had declined to 23 percent while the Chinese share had risen to 14 percent. Total Japanese trade was almost equal for the United States and China in 2006, at $213.7 billion and $211.4 billion, respectively, and in 2007 China passed the United States to become Japan's number one trading partner.

Table 6-1
East Asian Exports and Imports of Merchandise Trade
($ billions)

		Percent Share of Listed Economies	
	2006	**1995**	**2006**
Exports			
China	969	15	33
Japan	650	37	22
South Korea	326	11	11
Taiwan	224	10	8
ASEAN (10)	770	27	26
Imports			
China	792	17	31
Japan	580	30	23
South Korea	309	12	12
Taiwan	203	9	8
ASEAN (10)	685	32	27

Source: WTO, *International Trade Statistics*, 2006, Table III.69.

A similar rise of China to displace the United States and Japan as principal trading partners has been happening for South Korea and Taiwan, while the Chinese share of the Association of Southeast Asian Nations (ASEAN) trade grows steadily. The dominant Chinese economic power position within East Asia, moreover, carries over from trade to investment and finance. The $60 billion or more annual flow of FDI into China far exceeds the flow into others in the region, while the Chinese outflow of FDI is growing, particularly to Southeast Asia. In finance, the official Chinese foreign exchange holdings of $1.5 trillion by the end of 2007, headed for $2 trillion in 2008, provides unbounded financial resources for China to finance regional trade, outward investment, and economic assistance. Finally, the large size and high growth of the Chinese market makes China the principal macroeconomic influence for trade and investment flows within the region, a position which will grow even faster as the yuan is revalued.

Table 6-2
Japanese Merchandise Trade With Principal Trading Partners

	United States	**China**	**EU (25)***
Exports			
2006 Value ($ billions)	145.6	92.9	93.8
2006 Percent of total	23	14	15
1995 Percent of total	28	8	16
Imports			
2006 Value ($ billions)	68.1	118.5	59.8
2006 Percent of total	12	21	10
1995 Percent of total	23	11	16

*EU (15) for 1995.
Source: Japanese External Trade Organization, Trade and Investment Statistics.

Webster's defines hegemony as leadership, predominant influence, or domination, by one nation over others. China is now clearly the economic hegemon in East Asia in terms of predominant influence, while its leadership role is growing. The Chinese regional geopolitical objective is to build an East Asian economic grouping, through free trade agreements and financial linkages, with China as the dominant center. China negotiated a free trade agreement with ASEAN in 2005, for implementation by 2012, and supports an ASEAN plus three free trade arrangement to include Japan and South Korea. The Chiang Mai Initiative of 2000 created a network of currency swap loans within the ASEAN plus three grouping. In May 2007, their finance ministers agreed to set up a regional reserve pool that each member would be able to access in times of crisis, in effect a supplement to or substitute for IMF loans, with China a principal lender.

The rise of China to become the economic hegemon in East Asia, and its pursuit of a preferential regional trade and financial arrangement with China at the center, are growing concerns for others in the region, reflecting historical relationships in which China was the "Middle Kingdom," exacting tribute from its weaker neighbors. China repeatedly rejects any such intent, including the use of the term hegemon, but this can be seen as protesting too much.

The most important external dimension of the Chinese quest for an East Asian grouping is the exclusion of the United States, which is part of the broader Chinese strategy to limit U.S. power and influence in the region. In this context, regional free trade agreements, excluding the United States, would accelerate the decline in the U.S. market share of trade and investment in East Asia.

The principal policy counter-response by others in the region, and by the United States, is negotiation of free trade agreements across the Pacific, the Asia-Pacific free trade objective, as an alternative to an East Asian grouping that excludes the United States. The U.S.-South Korea free trade agreement of 2007, if approved by the U.S. Congress, would be the most important trans-Pacific agreement, and could lead to a U.S.-Japan agreement, which, in turn, would be decisive for the Asia-Pacific alternative. The subject of free trade agreements within East Asia and across the Pacific is discussed in detail in Chapter 9.

The most recent geopolitical development within Asia of great potential importance is the rise of India to become a major trading nation, with high priority on deepening economic ties within Asia. China has resisted India's inclusion within the East Asian grouping, for example by opposing full participation by India in East Asian summit meetings, but momentum is going India's way.

The Broadening and Balancing Economic Roles of India

India's rise as an emerging advanced technology superstate changes dramatically longer term geopolitical relationships within Asia. In terms of geographic scope, the "East" will drop out of the long-standing regional fixation on "East Asia." Asia will increasingly involve East and South Asia as a single integrated region. The incipient central economic characteristic of the region will be the bipolar relationship between China and India as two advanced technology superstates, each with over one billion population, the largest high-growth domestic markets, and the highest concentration of engineers, R&D, and multinational company and financial institution engagement. China is now far larger than India in terms of output and trade, but, as assessed in Chapter 5, the gap will progressively narrow over the coming 5 to 20 years.

This broadened and more balanced set of economic relationships within Asia as a result of India's rise has profound geopolitical implications which, on balance, should be highly positive for all, including for China. Deepening trade and investment with India would bring increased gains from trade for all. The greater geopolitical balance would reduce concern about China as the regional economic hegemon. In effect, Asia would move toward a dual hegemony regional economy, perhaps in need of new terminology, such as a "bihegemony," offering analytic scope for Ph.D. dissertations.

India clearly wants to be a principal part of this broader Asian economic relationship. It became a full participant at East Asian summit meetings, despite initial Chinese opposition. The 2004 India-Singapore comprehensive free trade and investment agreement was described earlier, and is viewed as a path-breaking success by both countries. A broader free trade negotiation between India and the ASEAN grouping got off to a slow start, because India presented a much longer list of "exceptions" to free trade than ASEAN did, but in January 2007 India reduced its exceptions list to less than the ASEAN list, at least in terms of the number of exceptions, and a final free trade accord could be concluded in 2008 or 2009. Another possibility is an Indian free trade agreement with South Korea in the wake of the U.S.-South Korea agreement.

There will also be geopolitical rivalries between authoritarian rule in China, together with Vietnam and Myanmar (Burma), and mostly democratic governments elsewhere, including the world's largest democracy, India. The China-India bilateral relationship will take on increasing importance over time, as will U.S. relationships with each. These relationships will involve commercial, foreign policy, and geo-strategic components. The geostrategic components will focus on China's rapid rise to become the dominant military power in Asia and the second global military power after the United States, and on the course of military modernization in India.

China as the Dominant Military Power in Asia

The rapid development of advanced technology industry in China has proceeded in parallel with a fundamental restructuring of Chinese defense industry, including integration of the civilian and defense industry sectors. Until 2007, the scope and pace of Chinese military modernization had been understated by the U.S. Department of Defense (DOD), as explained below. There is now near consensus, however, that China is the dominant military power within Asia and will at some point become the number two global military power after the United States. It is not clear when the asymmetric force structures of the United States and China will reach a balanced threat capability, particularly as related to Taiwan, but the U.S. lead is narrowing, and such balance may already be very close in some areas. This rise of Chinese military capability, moreover, has major geostrategic implications, not only for the United States, but also for all Asian nations, including India.

The performance capability of the Chinese People's Liberation Army (PLA) from the 1970s through the 1990s was dismal. In 1979, China launched a punitive attack on Vietnam and suffered enormous casualties against the smaller but more experienced Vietnamese defenders. The U.S. use of surgical bombing and electro-magnetic warfare in the 1991 Gulf War dramatically demonstrated the huge gap that China faced vis-à-vis the United States in modern weapon systems, a gap that was further displayed in 1996 when two U.S. aircraft carrier battle groups off the coast of Taiwan upstaged Chinese missile exercises with flight combat maneuvers and the monitoring of PLA activities on the ground.

The reasons for the PLA's failure to develop combat readiness and modern weapon systems have been analyzed extensively.[104] A continuing series of "reforms" within the military establishment were frustrated by vested interests in the status quo, a lack of incentives to improve performance, and the general isolation of highly secret defense facilities, even from one another. R&D institutes were separated from manufacturing facilities, preventing cost-benefit analysis at the development stage related to production costs and the weapons' ultimate performance. Corruption was also massive, stemming mainly from the large-scale production by defense industry enterprises of goods intended for commercial markets with weak accounting procedures.

Chinese leadership finally came to understand these failures, and from 1997 to 1999, a fundamental restructuring of the Chinese defense industry was adopted, closely linked, in concept and application, to the new framework for the market-oriented development of advanced technology civilian industry, launched a couple of years earlier. The National Defense Law of 1997 essentially subordinated the armed forces to state or civilian control, including, most importantly, for fiscal appropriations. The 1998 National Defense White Paper elaborated on this shift in control, giving the State Council responsibility for deciding

---

[104] See, for example, James C. Maulvenon and Richard H. Yang, eds., *The People's Liberation Army in the Information Age* (Rand Corporation, 1998); Larry M. Wortzel, ed., *The Chinese Armed Forces in the 21st Century*, (Strategic Studies Institute, 1999; and David Shambaugh, *Modernizing China's Military: Progress, Problems, and Prospects* (University of California Press, 2002).

the size, structure, and location of defense assets. With regard to the defense industry, in particular, the Council implemented three basic changes: it shifted control of the very large state-owned defense enterprises from the military to the civilian government; it integrated these defense industry enterprises more fully with other advanced technology enterprises for weapons development, including through joint R&D programs at universities and elsewhere; and it subjected defense projects to competitive bidding among defense and other enterprises, based on price and performance. The divestiture of the PLA's commercial operations took place during 1998 and 1999.[105]

This fundamental restructuring of the Chinese defense industry constituted, in effect, a rejection of the failed Soviet model, in which military facilities operated by administrative decree in isolation from the rest of the economy, and movement toward the U.S. model of civilian defense companies competing on the basis of price and performance, with considerable interaction between primary defense contractors and many other companies engaged in everything from R&D to dual use components.

The defense industry reform decisions of 1997 to 1999 initially met with skepticism, if not dismissal, by foreign observers. This reaction was based largely on the consistent failure of previous PLA reforms, and the cautious view that several years of credible implementation would be required before a positive assessment could be made. A commentary from 1999 concluded, "Whereas the PLA's ambitions were clear, the gap between ambition and capability could well be growing . . . what should be anticipated is a slow and sometimes erratic expansion of CMIC [Chinese Military Industrial Complex] capabilities."[106] A study in 2002 reached a similar conclusion: "Although the PLA has embarked on a systematic and extensive modernization program . . . a combination of domestic handicaps and foreign constraints severely limits both the pace and the scope of China's military progress."[107]

It was not until 2004 that a path-breaking paper by Evan Medeiros of the Rand Corporation laid out the changed circumstances:

---

[105] See Tai Ming Cheung, *China's Entrepreneurial Army*, chapter 10, "The PLA's Divestiture From Business, 1998-1999" (Oxford University Press, 2001).

[106] Bernard D. Cole and Paul H. B. Godwin, "Advanced Military Technology of the PLA: Priorities and Capabilities for the 21st Century," contained in Wortzel, *op. cit.*, 1999, pp. 209-210.

[107] Shambaugh, *op. cit.*, 2002, p. 10.

> In the last five years, China's defense industry has become far more productive than in past decades. The defense industrial reforms implemented in the late 1990s, unlike the ones adopted in previous years, were substantial and have positively influenced the quality of China's defense industry output. . . . There are a growing number of firms that do not belong to the eleven defense-industrial conglomerates (especially in the information technology (IT) sector) which produce goods under contract for the military. The line between defense industrial firms and civilian firms in China is increasingly blurred. . . . In the last two years alone, Chinese defense factories have produced a variety of new weapons systems based on novel Chinese designs. Many are highly capable weapons platforms. The development of these weapons importantly reflects improvements in R&D techniques, design methods and production processes, especially compared to the 1980s and the 1990s.[108]

Recognition of Chinese defense industry reforms and resulting weapons modernization was especially slow by the DOD, as reflected in its annual reports to the Congress on Chinese military power. The slowness resulted in large part from differences of opinion within the department between career intelligence China specialists tending to have a more cautious and benign assessment of Chinese capability and the senior leadership.[109] During the 1990s, the DOD reports had assessed the Chinese military capability to be 20 years or more behind the United States. The 2004 report, several months after the Medeiros presentation quoted above, devoted only 1 out of 54 pages to "domestic defense industry," and the conclusion was that "Chinese defense industries had taken near-term steps to address deficiencies, but Beijing realizes that long-term modernization will take time and entail a variety of measures . . . with few exceptions, such as ballistic missile research, development, and production, most of China's domestic defense industries are inefficient and remain vulnerable to dependencies on foreign suppliers of technology."

---

[108] Evan S. Medeiros, "Analyzing China's Defense Industries and the Implications for Chinese Military Modernization" (Paper presented before the U.S.-China Economic and Security Review Commission, February 6, 2004). The quotes are from pp. 1, 3, and 9.

[109] Department of Defense, *Annual Report on the Military Power of the People's Republic of China*, at www.defenselink.mil/pubs.

The 2005 report, delayed several months because of internal differences over the assessment, contained far more detail on advanced weapon systems under development and production, but with contradictory conclusions: "PLA modernization has accelerated since the mid to late 1990s . . . China rolled out several new weapon systems where development was not previously known in the West" (page 16), versus "China has not yet demonstrated the ability for innovation to go through a research, development and acquisition process for a sophisticated weapon system without foreign assistance" (page 24). The 2006 report continued to indicate differing views, although they took the form of papered-over conclusions rather than explicit contradiction. The continued "surprises" emerging from Chinese defense industry had been particularly vexing, as expressed by Assistant Secretary of Defense for International Security Affairs Peter Rodman:

> The widespread use of denial and deception at the strategic, operational, and tactical levels makes gaining "ground truth" on Chinese security affairs a difficult undertaking. We have been surprised in the past and we should expect to be surprised in the future. The summer 2004 appearance of the yuan-class submarine exemplifies this, as we had no prior knowledge of its existence. We have also been surprised at the pace and scope of China's strategic forces modernization.[110]

There is no doubt about the Chinese denial and deception strategy. It is rooted in the oft-quoted statement from Deng Xiaoping, "Hide our capacities and bide our time."[111] It also dates back centuries in Chinese strategic thinking, to Sun Zi's *Art of War,* and the "Shi" concept of strategic power, described as follows: "The lynchpin of the *shi* approach to strategy is superior information, an intelligence advantage. . . . When we look at the record of Chinese behavior we should see 1) a willingness to make abrupt shifts; 2) an emphasis on information dominance and deception; and 3) plans designed to undermine enemies at the outset of hostilities."[112]

---

[110] In a statement before the U.S.-China Economic and Security Review Commission, March 16, 2006.

[111] Quoted as the epigraph for the introductory chapter of the 2005 DOD report.

[112] Jacqueline Newmyer, in a statement before the U.S.-China Economic and Security Review Commission, March 16, 2006.

The May 2007 DOD report contained more detail on new Chinese weapon systems and a much higher grade assessment: "China's military transformation has increased in recent years, fueled by continued high rates of investment in its domestic defense and science and technology industries, acquisition of advanced foreign weapons, and far reaching reforms of the armed forces." Antisatellite weapons, mobile land-based intercontinental missiles, and new nuclear submarines with long-range missiles were highlighted. Chinese "informatization," or cyber warfare, capability was emphasized, with reference to information warfare units that would develop viruses with which to attack enemy computer systems and networks. With respect to Taiwan, "the balance of forces [is] continuing to shift in the mainland's favor." Chinese naval objectives were assessed as extending to the Malacca Straits and beyond, and a senior Chinese general was quoted: "Aircraft carriers are indispensable if we want to protect our interests in oceans." China is reducing its dependence on foreign assistance, with arms agreements in 2005 at only $2.8 billion, while in the long term: "Beijing seeks a wholly indigenous defense industrial sector."[113] This report stands in stark and sobering contrast with the previous reports described above.

An assessment of Chinese military capability, and Chinese strategy of denial and deception, extends to the estimated overall level of defense expenditures. A major shortcoming of DOD estimates is the estimated overall level of defense expenditures, through the exclusive use of the exchange rate measure for expenditures, which greatly understates the quantities of comparable Chinese weapon systems and military personnel.[114] The appropriate measure, as explained in the annex of Chapter 1, is the purchasing power parity (PPP) measure, which for Chinese GDP is estimated to be two and a half times larger than that based on the exchange rate measure. China, of course, never refers to the PPP measure for GDP, R&D, or defense expenditure comparisons, in keeping with Deng's rule of hiding capacity, which carries over to economic modernization generally, wherein China claims to be a poor, developing country. But why the DOD ignores the

---

[113] See www.defenselink.mil/pubs/pdfs/070523-china-military-power-final.pdf. The quotes are from pp. I, 5, and 24.

[114] The 2007 DOD report, for the first time, makes reference to independent PPP estimates for 2003, which indicated Chinese comparative spending two to three times higher than based on the exchange rate measure. The 2007 report itself, however, provides estimates of China's total military-related spending based solely on the exchange rate measure, without qualification.

appropriate PPP measure for Chinese defense expenditures is puzzling to say the least. Costs are clearly far lower in China for domestically produced comparable weapon systems and military personnel. China has the most modern shipyards and can produce a comparable submarine at a fraction of the U.S. cost, while the relative cost of a submarine crew is likely to be an even smaller fraction of the U.S. cost. Similarly for missile systems, defense R&D programs, and military and defense industry personnel across the board.

PPP figures are estimates, and the DOD should at least estimate the cost differential for major components of the Chinese defense budget. In any event, the Keynesian advice is critical that it is better to be approximately right than precisely wrong. In 2007, official Chinese defense budget expenditures, based on the exchange rate measure, were about $45 billion, up 18 percent from 2006. The 2007 DOD report estimated that additional off budget and hidden expenditures raised the total to a range of from $85 billion to $125 billion, which was 19 to 28 percent of U.S. defense expenditures of $439 billion,[115] all based on the exchange rate measure. With the PPP measure, in contrast, using the two and a half to one estimate related to GDP, Chinese defense expenditures would rise to the range of from $210 billion to $315 billion, or 48 to 72 percent of the U.S. level. In any event, the exchange rate measure greatly understates Chinese defense expenditures compared with U.S. expenditures, and this should be explicitly stated in the DOD annual reports, with at least some indications of how large the understatement is.

A detailed assessment of Chinese weapon systems is not provided here. The presentation is limited to a commentary on the Chinese navy because of its potential major interaction with the Indian and U.S. navies, of particular relevance to this study. Of the armed services, the navy has been out front in the development of advanced weapons and vessels. The navy receives high priority in terms of modernizing force structure and is the most deeply integrated with its civilian counterpart industry. China is second to South Korea in shipbuilding, and has leading-edge shipbuilding capability. In 2004, the Huadong shipyard contracted to build five advanced-design liquid natural gas carriers. The military component of shipbuilding includes ever more advanced design submarines, destroyers, and patrol craft. The success of the submarine program was highlighted in October 2006, when an untracked long-distance Chinese submarine surprised a U.S. aircraft carrier battle group off the coast of Japan, by suddenly surfacing and submerging. In December 2006, President Hu urged the navy to continue to build a

[115] The U.S. figure does not include supplementals related to the Iraq war.

blue water or deep sea fleet that can adapt to the navy's historic mission.[116]

Such blue water fleet operations would presumably extend to the Malacca Straits and beyond to protect vital Chinese oil imports from the Middle East. Other naval objectives with considerable "historic mission" context relate to land and sea transportation routes for the trade of the Chinese inland, western provinces. China has provided large military and economic assistance to the military government in Myanmar, including for deep water port construction and connecting roads from China to the Bay of Bengal. China also financed 80 percent of the $250 million construction costs of a deep water port in Gwadar, Pakistan, which could serve to anchor pipelines for oil and gas supplies from the Middle East to western China. There is a strong presumption that the Chinese navy would have docking privileges in these ports, as well as, perhaps, a port in Bangladesh.

A major question pertains to Chinese aircraft carrier fleet development. In 2002, China purchased an uncompleted aircraft carrier from Russia, with the deceptive story that it would be turned into a casino. It has since been taken into hiding and is believed to be subject to reverse engineering and activation as a training deck for aircraft and pilots. In March 2006, a Chinese general stated that in three to five years China will conduct research and build an aircraft carrier and develop its own aircraft carrier fleet.[117] The issue is thus no longer whether China will build a carrier fleet, but when and how.

An overall assessment of Chinese military capability relative to the United States is beyond the scope of this study. There should be no question, however, that the 20-year technological lead assessed by the DOD less than a decade ago has been greatly reduced and that in some areas the balance is close to parity. China is not only the dominant military power in Asia, but now the second global military power after the United States in the Asia-Pacific region, striving successfully, in terms of the definition of an advanced technology superstate, to become a military superpower.

Indian Military Modernization

Indian geostrategic interests and military force structure center heavily on relationships with its two largest neighbors, Pakistan and

---

[116] See China Reform Monitor, No. 653, February 6, 2007.
[117] From Richard D. Fisher, Jr., in a statement before the U.S.-China Economic and Security Review Commission, March 16, 2006, p. 6.

China. Wars were fought with both countries during the lifetime of the current Indian leadership, and the problems that caused those wars remain largely unresolved. India also has growing aspirations to become a global military power. Consequently, a broadly-based military modernization program is taking shape in India, with two driving forces, both directly related to this study. The first is the recent rise of civilian advanced technology industry in India, which provides a growing capability, as it does in China, for the modernization of the Indian defense industry. And the second driving force is the rapid modernization of the Chinese defense industry and weapon systems capability since the late 1990s, which has created strong political and emotional pressures in India to upgrade its armed forces so as to maintain a reasonable degree of balance in bilateral military capability.

A full assessment of the Indian military prospect is again beyond the scope of this study. The presentation here is limited to a brief description of the current state of the two neighboring bilateral relationships, followed by commentary on three recent developments that are causing substantial change in India's defense and geostrategic orientation: the modernization of Indian defense industry, the India-China blue water fleet relationship, and the deepening Indian defense relationship with the United States.

India had two post-independence wars with Pakistan, one in 1965 over Kashmir and the other in 1972 over Bangladeshi independence. India was the military victor in both wars, although the status quo ante was maintained in Kashmir, and violence and potential armed conflict remain a threat to the unresolved status of the Indian state of Kashmir. More recent bilateral tension involves the rise in radical Muslim terrorism and Indian accusations that Pakistan condones, if not supports, such terrorist acts in India. Finally, there are the bilaterally-oriented nuclear weapons capabilities in both countries, which are a two-edged sword. The "mutually assured destruction," or MAD, nuclear relationship in South Asia, like the U.S.-Soviet relationship in the Cold War, acts as a restraint on a third conventional war. At the same time, however, there is the threat of an unintended nuclear attack, especially as delivery missile capabilities grow in circumstances where bilateral command and control safeguards are more vulnerable than were the U.S.-Soviet safeguards. The Indian film industry, "Bollywood," could do a public service by demonstrating this frightening prospect through a remake of "Dr. Strangelove," relocated in contemporary South Asia.

China crushed the Indian army in the 1962 war over territorial disputes in two border states, and the disputes remain unresolved. President Hu's visit to India in November 2006, billed as the "Year of

Friendship," was marred when the Chinese ambassador to India told the Indian press, "The whole of what you call the state of Arunachal Pradesh is Chinese territory . . . we are claiming all of that—that's our position,"[118] and the Indian Foreign Minister Pranab Mukherjee responded that it was an "integral and inalienable part of India."[119] India has also been long troubled about Chinese assistance for the Pakistani nuclear and missile programs. A more recent and growing Indian concern is the Chinese intent to establish a blue water fleet presence in the Bay of Bengal, as described above. In broadest terms, India views itself, in historical terms, as the dominant military power in South Asia, a position now challenged by the rapid modernization and extension of the Chinese military.

The Modernization of Indian Defense Industry

In 2005, India had the third largest armed forces, after China and the United States, as shown in Table 6-3. Indian armed forces, at 1.3 million, moreover, were the only ones that have grown in size since 1990, whereas those of China, the United States, and Russia are down sharply.[120] India also became a nuclear power in 1998. In terms of conventional force structure, however, India lags far behind the other three countries. Indian military expenditures are only about one-quarter those of China, while 70 percent of equipment expenditures have been for imported weapon systems, sometimes outdated models. In quantitative terms, the weapons inventory is also much smaller than those of China, the United States, and Russia.

Indian defense industry has suffered from poor performance for many of the same reasons that plagued the Chinese industry. State-owned defense companies, overmanned and inefficient, operate in isolation from the rest of the economy, with inadequate checks on cost-effectiveness and performance and vulnerability to corruption. Earlier Indian aircraft projects have been characterized as "over-ambitious failures," with underpowered dangerous aircraft leading to frequent crashes and the deaths of test pilots. In any event, critical components had to be imported, mainly from the Soviet Union, and there was little or no transfer of Soviet design or manufacturing technology to India.

---

[118] The 2007 DOD Report, p. 8.
[119] *Financial Times*, November 20, 2006.
[120] The next two largest armed forces in 2005 were South Korea at 688,000 and Pakistan at 619,000.

India's lone success story was in the area of missile development and related space launch vehicles and satellites.[121]

Table 6-3
The Largest Armed Forces
(thousands)

	1990	**1995**	**2005**
China	3,030	2,930	2,255
United States	2,118	1,595	1,474
India	1,262	1,145	1,325
North Korea	1,111	1,128	1,106
Russia	3,988	1,520	1,037

Source: International Institute of Strategic Studies, *The Military Balance*, various years.

This is now changing fairly rapidly, as India insists on co-production with foreign arms suppliers, while domestic defense contracts are being progressively opened to private companies. In 2005, a government decision called for 30 percent of foreign defense contracts above $66 million to be offset by purchases, investments, and transfer of technology in India. For joint venture projects within India, the normal limitation of 26 percent for the foreign partner is being relaxed. For example, a 50-50 $11 million joint venture between Snecma of France and Hindustan Aeronautics was approved in the aeronautical sector.

A very large, ambitious project is the 2005 $3 billion joint venture between the Mazagon Dock shipyard in Mumbai and Armaris, a subsidiary of the French Thales Group. Phase one is to build six Scorpene diesel attack submarines. Construction began in December 2006, with one submarine per year to be launched beginning in 2012. The contract includes the transfer of technology, with just the first two submarines being built under French and Spanish supervision. Phase two is to build 18 additional submarines. This project has been marred, however, by allegations of corruption.

As for Indian private companies, the government opened defense contracts to them in 2001, although implementation got off to a slow

[121] See Stephen Philip Cohen, *India: Emerging Power* (Brookings Institution, 2001), especially chapter 4, "The Domestic Dimension." The quote is from page 99.

start. By early 2006, however, 19 companies had received 28 manufacturing licenses for supply contracts, and in March 2006, for the first time, the government awarded two $20 million contracts to private companies as prime contractors, to Tata Power and Larsen & Toubro, for design work for launchers in the Pinaka Missile System.

In parallel, Indian arms exports are growing, with the intent of making India a major arms exporter. Rahul Chaudhry, chief executive of Tata Power's strategic electronics division, commented: "Like Israel, India has the right skills and low cost design and testing capabilities to compete internationally."[122] An Indian-Russian joint project produced the BrahMos long-range supersonic cruise missile, shown in both countries to potential customers. Chile has expressed interest in the Indian made ALH Dhruv helicopter, produced in India with assistance from Israel.

Yet another dimension of what is, in effect, defense service exports is military training. India's long-standing Counter-Insurgency and Jungle Warfare School (CIJW) in Mizoram is one of the best anti-terrorism schools, catering to foreign military personnel, including American, and the school has a large backlog of requests for training from Western nations.[123]

All of these defense industry initiatives are at an early stage and, together, are of relatively modest scope. The bulk of Indian expenditures for defense equipment still go for foreign purchases, while domestic contracts almost all go to low-performance state-owned companies. The change toward greater participation by advanced technology private companies is under way, however, and will likely gain momentum over time. A central route for such change, again following the Chinese and American models, would be to open broadly defense contracts to competitive bidding, based on price and performance, from both public and private companies.

Indian and Chinese Blue Water Fleets

Barring the highly unlikely outbreak of war between India and either Pakistan or China, the most important emerging military interaction for India over the coming five to ten years will be between

---

[122] *The Economist*, April 8, 2006, p. 43.

[123] See Erica Lee Nelson, "Indian Military Gaining Global Prestige," *Washington Times*, July 29, 2006. This piece includes the references to the BrahMos missiles and the ALH Dhruv helicopter.

the Indian and Chinese navies in the Bay of Bengal down through the Malacca Straits. The Indian fleet consists of more than 150 vessels, soon to include a second aircraft carrier battle group. India has long considered itself a naval power of global reach, and certainly the dominant naval power in the Bay of Bengal as well as the superior naval power vis-à-vis Pakistan in the Arabian Sea. In the aftermath of the 2004 tsunami disaster in Indonesia, the Indian navy deployed 27 vessels and 25 helicopters and aircraft for earthquake relief, which was given prominent media invidious comparison with the near absence of Chinese relief efforts.

China, however, as recounted earlier, now plans to build a blue water fleet, including aircraft carrier battle groups, to protect its trade routes through the Malacca Straits and to extend its presence into the Bay of Bengal, and this is viewed with deep concern in India. Naval modernization, compared with the other armed services, is moving forward faster in both countries largely because of deeper integration of naval defense and civilian industries, mainly modern, internationally competitive shipyards. China is far ahead in shipbuilding, with 60 shipyards, and is currently building the world's largest one near Shanghai. India, however, is projecting a fivefold increase in ship construction by 2018, to $18 billion, and has two major new shipyards in the works.[124]

China is well ahead of India in military ship design and construction for submarines, including nuclear, destroyers, and patrol craft, but India is moving forward to achieve indigenous warship construction through foreign participation, as in the Scorpene joint venture submarine project. The highest visibility attention, however, will be for aircraft carrier construction, still in the planning stage in China, while already under way in India. India has one aging, British-built carrier, the Viraat, and purchased a Russian carrier in 2004, which has been renamed the Vikrama, and will be in operation by 2008. In addition, construction began in 2005 at the Cochin shipyard in India for a third aircraft carrier, to be launched in 2012. The objective is to have two carrier groups deployed at all times, with the third in maintenance and repair. Since the Viraat is scheduled to be retired by about 2014, the implication is that a second carrier will be built in India over the coming decade.

However the emerging Indian and Chinese blue water fleets take shape, the strategic questions concern how they will be deployed and for what purposes. For the Malacca Straits and sea lanes generally,

---

[124] See *Times of India*, May 11, 2007.

there are obvious mutual interests served by cooperative interaction to protect the waterways from pirates, terrorists, and storms. But it is not clear to what extent such cooperation will develop, in view of the extreme secrecy of Chinese military operations and the deepening concern of India about the deployment into the Bay of Bengal of advanced-design, long-range Chinese submarines and missile-carrying destroyers, not to mention aircraft carriers.

The presumed intent of China to obtain military basing rights in Myanmar brings the Chinese presence closer to home for India and raises even deeper concerns. For centuries, China has sought influence if not control over what is today Myanmar, as a land and sea route for its southwestern provinces, and this objective is now close to being achieved. As noted above, the beleaguered military government in Yangon, isolated from the West through economic sanctions, has become highly dependent on Chinese military and economic assistance. India now is also providing military assistance to Myanmar to counter Chinese influence, and in January 2007 Indian Foreign Minister Pranab Mukherjee visited Myanmar and said that India would grant a favorable response for new arms requests from Myanmar. In October, both China and India held back from global condemnation of the Myanmar military killing and suppression of democracy protestors led by Buddhist monks.

It is not clear how these relationships will develop, but there is no question that the Indian-Chinese naval relationship will broaden and deepen over the coming five to ten years. This engagement will also be of concern to others in the region, extending to Japan and other East Asians with respect to the Malacca Straits. The ASEAN grouping, troubled by member Myanmar's human rights abuses and lack of democracy, is faced with China and India as the two foreign nations with the largest official presence and influence within that country. And there is finally the United States, the third and most powerful blue water fleet in the Pacific, whose intentions further west toward the Bay of Bengal are unclear.

The Deepening U.S.-Indian Military Relationship

U.S.-Indian relations have been deepening on a positive track over the past several years, principally through trade and investment, but in broader foreign policy terms as well, as the succeeding chapters discuss. First steps have also been taken toward a more engaged military relationship, however, and these developments fit best here, at the conclusion of the discussion of Indian geostrategic interests.

The U.S.-Indian military relationship has been relatively distant for the past 50 years, as India depended primarily on Russian military equipment and related training, within a "nonaligned" foreign policy that, in military terms, was far more nonaligned to the East than to the West. This is now changing, however, and defense-related contacts and projects between India and the United States are on the rise, with considerable potential for becoming stronger. This stems largely from a mutual interest in maintaining a balance with the growing Chinese military capability in the region, although such self-evident mutual interest is not stated publicly.

In terms of Indian military procurement from the United States, some specific steps have already been taken. In 2006, India purchased the USS Trenton, renamed the Jalashva, an amphibious transport warship, together with six helicopters that will operate from the ship, for $50 million. The United States, in response to an Indian request, has also offered to sell F-16 and F/A-18 Hornet combat aircraft to India. These U.S. planes, however, are competing with the Russian MiG-35 and the French Mirage 2000-9, and many Indian air force officers are reported to favor the Russian and French companies as reliable past suppliers, including for training, transfers of technology and license production in India.

Looking ahead, three important dimensions of a strengthened U.S.-Indian military relationship will likely be:

Deepening institutional relationships. The importance of direct contacts between military institutions cannot be overstated. Professional pride and mutual respect run deep among military officers everywhere. There has been relatively little such interaction between the U.S. and Indian military services, but personal contacts are now expanding. They should lead to especially strong professional bonding among representatives of leading military powers who have shared values, while speaking a common language.

Defense industry relationships. Substantial business could develop between the two defense industries, as India progressively opens its military procurement to foreign companies. There is a high degree of complementarity between U.S. advanced weapon design and production capability and lower cost, increasingly high-tech Indian industry. The result could be production sharing between the two defense industries and acquisition by both militaries. An important U.S. policy issue in this regard would be the relaxation of strategic export and investment controls for India.

The three blue water fleet relationship. The future course of interaction among the U.S., Chinese, and Indian fleets in the Pacific Ocean, extended to the Malacca Straits and beyond, is not yet well defined but is beginning to take shape. The U.S. and Chinese navies have long been engaged adversarially in the Pacific, principally with respect to Taiwan, but the broader reach of the Chinese navy is now beginning to be discussed bilaterally. As for the U.S. and Indian navies, a landmark initiative was the joint exercises of the U.S., Indian, Japanese, and Australian navies in the Bay of Bengal in 2007, exercises that are likely to become an annual event. This was viewed negatively by China as a form of naval encirclement by the four democracies. The question is how such exercises will be explained in terms of objectives, and whether at some point they might be expanded to include China for the purpose of protecting vital trade routes. The only suggestion offered here is that an in-depth, forward-looking analysis of the options facing the regional navies be undertaken, preferably with joint participation. There should be no question, however, that over the medium to longer term, the relationship among the U.S., Chinese, and Indian navies will constitute a significant dimension of what can be called the new Asia-Pacific triangle.

PART II

The U.S. Policy Response

CHAPTER 7

The Rising New Asia-Pacific Triangle

The 2005 study concluded that China had become the most important U.S. bilateral relationship, that the economic relationship was paramount, and that wide-ranging national security and other international interests were far less engaged. The economic relationship, moreover, while providing large mutual gains from trade, was deeply troubled, over the currency manipulation and a number of trade policy issues, with the net result of a substantial loss of U.S. export competitiveness in manufactures and a serious threat to long-standing U.S. technological leadership. There was finally the ideological divide between American democracy and Chinese authoritarian rule, with adversarial consequences related to Chinese military expansion and U.S. support for continued democratic governance in Taiwan.

This study updates the assessment for the China relationship and extends the scope of analysis to include India within what is termed the new Asia-Pacific triangle. For China, all of the elements enumerated above have increased in force, mostly in an adverse direction, with the notable exception of U.S.-Chinese collaboration to terminate North Korea's nuclear weapons program. The new U.S.-India relationship, as described throughout the previous chapters, entails an equally wide-ranging set of issues, although with more convergence of interests and the beginnings of a much deeper positive relationship, even more concentrated in economic engagement than the U.S.-China relationship. As for the new three-way relationship, or Asia-Pacific triangle, it is still at an incipient stage and heading into largely uncharted policy waters, with the deepening ties among all three participants of the relationship thus far, once again, predominantly economic.

"Strategic triangles" have long intrigued political scientists and diplomatic historians, particularly to the extent that two of the three participants gang up on the third. Since 1950, principal attention has focused on the U.S.-China-Soviet and U.S.-Japan-China triangles, although the U.S.-China-India triangle has also received some attention. Harry Harding surveyed the evolution of the latter triangle since Indian independence and concluded that, until recently, it was a secondary pattern of international affairs because of India's economic weakness, and it never crystallized into a firm alignment of two against one.[125]

---

[125] Harry Harding, "The Evolution of the Strategic Triangle: China, India, and the United States," in *The India-China Relationship: What the United States Needs to Know*, ed., Francine R. Frankel and Harry Harding (Columbia University Press, 2004), pp. 321-350. The quotes in the next paragraph are from pages 338, 343, and 339.

Looking ahead, as of 2004, Harding predicted that "The triangular relationship may become more salient in the coming decades as the national power of India and China increases and as South Asia and East Asia are gradually integrated into a single region." As for the India-China relationship, "the most likely scenario is that of a competitive relationship, unbuffered by extensive economic ties, but without a significant possibility of armed conflict." Moreover, "the principal danger facing India is economic stagnation exacerbated by political gridlock."

What a difference three years make! The U.S.-China-India triangle is becoming more salient year by year rather than decade by decade, and the threat of Indian economic stagnation has turned into the reality of sustained very high growth and industrial modernization. This burgeoning new Asia-Pacific triangle—and the adjective "new" cannot be overemphasized—is far more deeply engaged in economic terms than it was only a few years ago, while the economic, national security, and foreign policy dimensions of the overall relationship are becoming progressively more interwoven.

The prospect for this Asia-Pacific trilateral relationship is momentous in content. It is on track to comprise the nations with the three largest populations, the three largest economies, the three largest military forces, and the three largest advanced technology infrastructures for new technology development and application. Geographically, these three advanced technology superstates will be the dominant economies within North America, East Asia, and South Asia, and together they will form the driving power structure for the entire Asia-Pacific region, which will account for 60 percent of the global population and considerably higher shares of global economic growth, military expenditures, and technological innovation.

This historic transformation calls for a forward-looking examination of U.S. interests and the formulation of a comprehensive policy response. The following presentation is designed to begin this examination in an orderly way, with the central focus on the economic relationships, which at this stage are the most deeply engaged and the most complex in terms of policy formulation and execution. As with the net analytic assessment in Chapter 5, the recommended policy response is far more specific for the short to medium term of two to five years. Chapters 8 through 10 address this time frame in terms of international finance, trade, and investment policies and the corresponding domestic policy agenda. The concluding Chapter 11 then provides a more broadly framed, longer term policy perspective.

Such economic policy formulation, however, cannot be done in isolation. The economic relationships are connected in many ways with broader foreign policy and national security interests. The previous chapter addressed much of this broader picture in terms of geopolitical and geostrategic issues. No attempt is made here at a comprehensive assessment of this new and evolving Asia-Pacific triangular relationship except to urge that such an updated assessment be undertaken. The introductory presentation which follows is limited to a brief discussion of three basic dimensions of the overall relationship as they interact with the advanced technology development theme of this study: the great ideological divide, national security interests, and international leadership roles. The chapter concludes with a recapitulation of the U.S. policy challenge ahead.

The Great Ideological Divide

The United States and India are the world's two largest democracies, while China is the predominant authoritarian power. This ideological divide underlies many of the national security and foreign policy differences between the United States and China. It also embodies the potential for a competitive if not adversarial political relationship between the two democracies and authoritarian China within the Asia-Pacific triangle. In contrast, to the extent China makes its officially intended transition to democratic governance, the three-way relationship would become far more closely aligned in political terms, and, consequently, the three would be more likely to act in concert across a wide range of international issues.

The most basic policy question for the United States is thus how to respond to the prospect for democratic change within China. The options range along a spectrum from "pragmatism," whereby the United States treats this as an internal Chinese matter except for established multilateral forums on human rights, to a forceful U.S. agenda to spur democratization in China, including the threat of economic sanctions, as was pursued during the late 1980s after the Tiananmen Square massacre and in the early 1990s related to most-favored-nation treatment for trade with China. Current policy is very close to the pragmatist end of the spectrum.

The policy response offered here is that the United States should be more proactive in support of democratization within China, but should pursue it principally in terms of public diplomacy and private sector engagement, and not through economic sanctions. Two developments within China support such a course. The first is that economic modernization in China, based on market-oriented reforms and economic freedom, has reached the stage where internal public pressures for democratization are growing and will force even the highly risk-averse current communist government to seriously consider genuine democratic reforms. As described earlier, the need for the rule of law and property rights in rural areas, the growing influence of a more affluent and educated middle class, and discontent over widespread corruption and nepotism all cry out for a more open and publicly accountable government. The immediate prospect of major structural adjustment, with uneven impact among industries and regions, together with mounting environmental problems, will require difficult decisions that will further increase pressures for public participation in such decisions. It is very likely that important political change will occur in China over the coming two to five years, which could be in the direction either of harsher authoritarian control or of the beginnings of democratic governance. The latter is clearly preferable for the Chinese people as well as for U.S. relations with China.

The second development, flowing from the first but less clear, is an apparent growing awareness within the communist government that the long-standing objective of democratization needs finally to be addressed in more serious and action-oriented form. It is significant that Premier Wen Jiabao, in March 2007, as a lead up to the Communist Party Congress in October, stated that democracy, the rule of law, freedom, and equality, were not something peculiar to capitalism but part of Chinese socialism as well. Cautious but pointed discussion of democratization was also building during 2007, without suppression. A former university vice president published an article concluding that "Not everything can be decided by one party or one person." Another magazine article attacked the notion that ordinary Chinese were not ready for democracy: "Democracy allows ordinary people to improve their own quality and that of the whole nation." The editor of the magazine commented that no complaints had been received about such articles.[126] At the same time, however, overall control of the media has been tightened.

The party leadership continues to state categorically that one party rule will continue, although with greater democratic process within the party. One party democracy, however, is the ultimate oxymoron, precluding all basic elements of democratic governance, including the rule of law through an independent judiciary, checks and balances on government power and corruption, individual rights, freedom of the press, and, of course, democratic elections. If the communist leadership is to take credible steps toward democratization, these fundamental issues need to be addressed.

The Party Congress in October 2007 produced a mixed and almost bizarre outcome for democratization. President Hu used the democracy term about 60 times in his two-hour speech, but presented it as a socialist alternative to Western democracy within one party rule. The central substantive theme of the meeting—greater gains for the rural poor—could imply their greater participation in the political process,

---

[126] See Richard McGregor, "China Opens Up to Redefine Democracy," *Financial Times*, June 13, 2007.

but the large increase in the share of the Central Committee provided to the Peoples Liberation Army indicated a different direction of power sharing. The net assessment here is that the issue of democratization is front and center in the minds of China's top leadership, as witnessed by the 60 references, but that no significant decisions have yet been made as to what to do about it, all within the context of growing concern by the leadership about challenges to their cherished one party rule.

In this setting, a more proactive U.S. public diplomacy to encourage democratization in China should be productive. The essential elements of democratic governance should be stated clearly and regularly, as should the fact that "one party democracy" is an oxymoron. It should also be emphasized that a democratic transition in China can and should be peaceful and non-disruptive, as it was in the Soviet Union, East Europe except for Romania, and, closer to home, South Korea and Taiwan. The specific steps for achieving the democratization objective are the domain of the Chinese government and people, but the United States should state that it would welcome dialogue on how the Chinese government plans to move forward to its stated democratization objective and how the United States might be supportive.

The modalities for pursuing more proactive U.S. support for democratization in China would involve both the public and private sectors. In the public sector, U.S. support for international bodies, nongovernmental organizations, and academic discussion dealing with human rights, the rule of law, and democratization more broadly would be strengthened. Criticism of violation of basic human rights in China would be more forceful and frequent. The United States could also offer support for political reforms, such as those strengthening the technical modalities for applying the rule of law. The highest visibility step to consider at some point would be a statement by the American president that the time had come to tear down this Chinese wall against individual freedom and democracy.

The democratization role of the U.S. private sector, which is deeply engaged on the ground in China, could be more important and is already being played. American companies in China reflect American values through the way they treat their employees, their respect for the rule of law, and the greater openness in their operations. This positive influence should be supported by the U.S. government, most importantly through a more forceful economic strategy to achieve a deeper and more balanced economic relationship with China, as presented in Chapters 8 through 10, which, among other things, would stave off a protectionist backlash in the United States and a reduced private sector presence in China. The U.S. government would also give greater public recognition to the value of having U.S. companies doing business in China. When China orders foreign companies to take actions that reinforce authoritarian rule, such as acquiring information to use in prosecuting political dissidents, the United States should protest loudly and clearly to the Chinese government, while acknowledging that American companies in China are basically obliged to comply with Chinese law.

The role of India in support for democracy and the rule of law will likely remain modest on the international front, and will be played out principally through the demonstration effect within India, where, as described earlier, democratic process is deepening and strengthening as a result of industrial modernization and the application of new information technologies. In multilateral forums, India, reflecting its anticolonial history, has been deeply committed to respecting national sovereignty and therefore wary of proposals to promote worldwide democracy. This should gradually change, however, as India acquires growing influence as a global power and stands out as the largest democracy. The United States should encourage collaboration in this direction, including in multilateral forums for the protection of basic human rights and freedom of the press and religion.

As for Indian-Chinese dialogue on the subject of democratization, there is no immediate prospect of such dialogue, but there is potential for useful technical exchanges to the extent China begins democratization in earnest. The two countries share the challenge of bringing the rule of law and individual rights to their large rural poor populations, and, once China begins to address this set of issues, technical exchanges with India would have fewer political overtones than would similar dialogue with the United States.

The subject of democratization in China and its impact on the course ahead for the three advanced technology superstates—whether working together in closer concert as the three largest democracies or in a two-to-one split with adversarial implications—is of momentous consequence for the future of the international economic and political orders. It should be addressed among the leaders of the three nations, with a view to convergence over time to democratic governance. How to undertake such dialogue, or trialogue, in a nonpoliticized yet serious manner is a daunting challenge for international statecraft. All three governments should give it their best try.

National Security Interests

The three nations have mutual national security interests for the nonproliferation of weapons of mass destruction (WMDs) and the suppression of international terrorism, although the perceived threats and commitments to action vary. There are also bilateral and regional national security interests in play that are adversarial, between the

United States and China and, to a lesser extent, between China and India.

In the field of WMDs, concerted action has been limited. For China, the most important initiative has been collaboration with the United States within the group of six to terminate the North Korean nuclear weapons program. China has the greatest leverage through its critical economic support for North Korea, although its interest in shutting down the nuclear program is tempered by its competing interests in avoiding regime change in the North and Korean reunification. For the Iranian nuclear program, China has been far less forthcoming and more prone to block UN pressures through economic sanctions against Iran. Chinese interests in Iranian oil and a reduced U.S. power position in the Middle East presumably outweigh Chinese concern about an Iranian nuclear weapons capability. U.S.-Chinese cooperation to limit the proliferation of biological weapons has also been very limited, with the Chinese role more to criticize inconsistencies in U.S. policy than to take positive steps of its own.

Indian collaboration on nonproliferation has also been limited, partly because of its own nuclear ambitions. With respect to Iran, India doesn't want another nuclear power in the region, but it has competing interests in Iranian oil and gas supplies and in access to central Asia through Iran, access which Pakistan refuses to grant. It was therefore significant that in 2007 India voted twice with the United States against Iran in the International Atomic Energy Agency, which was highly controversial within India. The outlook is for India to continue to work with the United States on the Iranian nuclear proliferation issue while at the same time developing a more open and deeper economic relationship with Iran.

Indian collaboration on nonproliferation is also linked to the U.S.-Indian civil nuclear agreement, in the uncertain process of being approved into 2008. India is highly sensitive to and rankled by the asymmetry whereby China is a legitimate nuclear power and India is not. If the agreement is concluded, along with the safeguard agreements with the International Atomic Energy Agency and the Nuclear Suppliers Group, India will have achieved not only economic support for its civilian nuclear power but also at least semilegitimacy as a nuclear power. For these reasons, China likely opposes the agreement and could play an obstructionist role in the Nuclear Suppliers Group.[127]

For antiterrorist collaboration, there is a more direct common interest, with each of the three nations facing an internal threat from radical Muslim terrorists. Cooperation is closer between the United

[127] Some in India believe that China put the Indian communist parties up to opposing the agreement within the coalition government.

States and India, however, because they confront the greater threats. China faces a relatively smaller internal threat and has mixed motives about the U.S. war on terror in Afghanistan, Iraq, and the Middle East more broadly. Reports of Chinese exports of weapons to Iran that are then reexported to Afghanistan and Iraq are particularly disturbing. In any event, the new Asia-Pacific triangle is not now a significant dimension of the global response to Muslim terrorist groups, although U.S.-Indian collaboration could strengthen.

The adversarial bilateral and regional national security interests have already been discussed in Chapter 6. For Taiwan, the United States and China have a strong mutual interest in avoiding a crisis leading to military or other hostile acts against Taiwan by China. The best solution is to maintain the status quo, which includes constraining Taiwan from declaring independence. It is unclear how the deepening economic interdependence between China and Taiwan will affect the U.S.-China relationship, although it clearly increases Chinese leverage over Taiwan. A critical factor is the course of democratization within China. Significant democratization within China would open the way for China to accept that any political change in Taiwan would be subject to prior approval by the Taiwanese people.

There is likewise a mutual interest between India and China to avoid confrontation over their border disputes, despite the hard-line public positions on both sides and a deep sense of historical grief in India over the 1962 war. The best course ahead would be for both sides to continue along the path of rapid economic modernization and to rise to become global economic powers, at which point both would be better positioned to compromise on what are largely remote and low-productivity border areas.

The looming and least clearly defined national security issue, which is headed toward an adversarial relationship between the United States and India vis-à-vis China, is the rapid expansion and modernization of the Chinese military, with extended outreach throughout the East and South Asian regions. Repeated U.S. requests to China to discuss Chinese military strategy and intentions have been resisted. Chinese-Indian military strategic dialogue is also minimal. The leading-edge issue of interrelationships among the three blue water fleets was discussed earlier.

The Chinese military buildup goes to the core issue of dominant power relationships in Asia, with a U.S.-Indian mutual interest in denying Chinese dominance and a corresponding Chinese interest in avoiding containment by the United States, India, and other Asian democracies. For India, it also relates to Chinese support for Pakistan, including for weapons development. The rise of China to become a global military power greatly raises the stakes for building a more cooperative three-way military relationship for regional peace and the protection of sea and air transport, although the immediate prospects are unpromising.

The principal observation from these various national security interests for the discussion of the economic relationship in subsequent chapters is that the national security interests are largely separate from and unrelated to the current predominant economic relationship, except where specific direct linkages are involved. One important direct linkage is between Chinese Middle East oil interests and the Iranian nuclear program. Another potentially critical linkage is the likelihood of severe U.S. trade and other economic sanctions against China if China were to take military or other actions to suppress democratic governance in Taiwan.

This line of assessment leads to the question of the extent to which there should be indirect trade-offs between national security and economic interests. The conclusion drawn here is that such trade-offs would be of little utility and should be avoided. Certainly the far broader U.S. commercial and other economic interests with China and India should not be sacrificed for limited and often dubious concessions in the national security domain. Mutual national security interests will only be advanced when it is in the national security self-interest of the participants to do so. This certainly applies, for example, to the U.S.-China collaboration to terminate the North Korean nuclear weapons program. Consequently, recurring suggestions to go easy on China in economic negotiations in order to obtain a more forthcoming Chinese position on North Korea or Iran should be rejected.

Foreign Policy Leadership Roles

The rise of China and India toward joining the United States as the three advanced technology superstates raises fundamental questions about their respective leadership roles in the global economic and political orders. Will they play constructive roles to strengthen multilateral norms and obligations, or will they pursue a more narrow course of national self-interest? Will they act more in concert or as rivals, if not adversaries? These questions are of great longer term historical consequence, as discussed in Chapter 11. The more immediate outlook is that they will play widely divergent roles, with relatively little concerted leadership. This is likely to begin to change, however, over the next several years. Opportunities will likely present themselves for more concerted and influential leadership, and the three governments can and should rise to the occasion.

The United States has played a major international leadership role for over a century and is currently the sole global superpower. It is forcefully engaged in almost all major international political and economic issues, generally in support of international norms and obligations, ranging from maintaining peace to respecting basic human rights to supporting free trade. Such leadership is deeply embedded in the national culture, although political pressures have been building to pursue a narrower course of national self-interest.

China, in contrast, largely pursues the national self-interest course, often taking positions that weaken international norms, and rarely that strengthen them. For example, it has played no significant leadership role in the troubled WTO Doha Round despite its rise to become the number one exporting nation with the largest trade surplus in history. China's objectives in Sudan, Zimbabwe, and other poor nations to obtain oil and other investment contracts for Chinese companies are often at odds with UN and other initiatives to reduce violence and the violation of basic human rights, if not genocide. China has begun to discuss a more cooperative international "stakeholder" role, but there has been little collaborative Chinese leadership, except for shutting down the North Korean nuclear weapons program, which in any event primarily involves a direct Chinese national interest.

India is distinct in that it has played an active international leadership role since independence, with great national pride, but has been a relatively weak power in global terms, and its main leadership objectives, such as pursued through the nonaligned movement and a highly protectionist, "developing country" position in the trading system have put it at odds with the United States.

The question is how these disparate leadership roles might change over the coming two to five years and what the United States can do to build more positive and concerted leadership among the three nations. This question is addressed in detail in the following chapters for the economic relationship, writ large to include such issues as energy and the environment. For the broader political order, including national security interests, comments have already been made about some areas of potential concerted action. Two broad issues stand out, however, as roadblocks to such concerted leadership. The first is the ideological divide between the United States and China, already discussed, reinforced by China's deepening nationalist foreign policy orientation, which tends to position the two countries as rivals, with divergent interests, rather than as partners in international leadership. The Chinese initiative to form a China-India-Russia grouping as a counterweight to U.S. unipolar influence is a manifestation of such distancing from the United States.

The second roadblock to more concerted leadership, particularly but not exclusively in the economic area, is the long-standing North/South dichotomy among nations, between "developed" and "developing" nations. This simplistic dichotomy is highly misleading and counterproductive because the community of nations spans a continuous spectrum from the most to the least developed, and there is no rational dividing line for separating all nations into just two fundamentally different categories. This misleading dichotomy is becoming increasingly divisive as a number of newly industrialized nations, and China most of all, rise to positions where they need to take on the full obligations of membership in international organizations if these organizations are to function effectively. China is clearly a developed country by many specific comparative standards, from international trade and finance to military might. India and a number of other newly industrialized nations are likewise moving up the developmental ladder and are much closer to the "developed" grouping in terms of economic and, in some cases, military power, than they are to the many "least developed" countries on the lower half of the ladder. This issue is discussed in detail in later chapters with respect to the international trade and financial systems and the evolving role of official development assistance, and is raised here to highlight the fact that concerted leadership among the United States, China, and India will be limited as long as the latter two maintain solidarity within a "developing country" grouping, in a largely adversarial relationship with a "developed country" grouping that includes the United States.

The general conclusion is that concerted international leadership among the United States, China, and India is limited and will evolve slowly. There are also important differences among the three bilateral relationships, with the U.S.-China relationship the most important and the most troubled, the U.S.-India relationship on much firmer ground and strengthening, and the China-India relationship a mixture of strategic competition, economic and political engagement, and wary watching. Closer collaboration among all three should nevertheless be possible and should be actively developed as opportunities present themselves. Energy and the environment are obvious sectoral candidates, although there are scant results to date. Moreover, as the three actual and emerging advanced technology superstates, the full range of new technology horizons could become fertile ground for tripartite joint leadership in international forums.

The U.S. Policy Challenge Ahead

The rising new Asia-Pacific triangle presents a new and far-reaching challenge for U.S. foreign policy. At this stage, however, relationships among the United States, China, and India are very distinct and only loosely connected, both in bilateral and multilateral terms and within the principal areas of policy. The U.S. political relationship with democratic India is on a highly positive track, and should strengthen as India rises to become a more self-confident, modern information-age democratic society. The U.S. policy response needs to be comprehensive in nurturing the many opportunities for building a stronger, more collaborative relationship. The U.S.-China political relationship, in contrast, is both far more important and far more troubled. The various national security and foreign policy issues engaged need to be addressed on an issue-by-issue basis, with a view to more collaborative achievement. The rapid expansion and modernization of the Chinese military and the underlying ideological divide, however, cast threatening clouds over the relationship.

This leaves the economic relationship, where the most intense and complex interactions are currently taking place, and to which the following three chapters are devoted. The foundation of the relationship is highly positive in terms of potentially very large, highly dynamic gains from trade, investment, and technology transfer. The high rates of growth in China and India, the second and fourth largest national economies, combined with the United States as the central driving force for new technology development and application, provide the setting for unprecedented mutual economic gains for each of the three as well as for the global economy.

Two immediate challenges, however, threaten the achievement of this potential and need to be addressed with forward-looking vigor. The first is the great imbalance in the Chinese economy toward mercantilist, export-oriented growth, together with the daunting internal obstacles to more domestically oriented Chinese growth. This trade restructuring challenge for China has broad geographic implications, but it focuses heavily on the U.S.-China bilateral imbalance. The second challenge is the unraveling of the international financial and trading systems which have prevailed over the past 60 years, and the urgent need to reestablish their effective functioning. In this case, the outcome is even more decisively centered on the new Asia-Pacific triangle: If the United States, China, and India can together agree on the needed changes, the changes will happen; and if they cannot agree, the changes will not happen.

CHAPTER 8

International Financial Policy

The most important and immediate U.S. conflict with China is in the area of international finance, the "currency manipulation" issue, whereby China is maintaining its exchange rate to the dollar far below a market-determined level, thus gaining an unfair trade advantage, with substantial adverse impact on U.S. exports and on advanced technology industries in particular. Ill-conceived attempts by the United States from 2001 to 2006 to deal with the issue got nowhere. In 2007, the currency issue finally became engaged multilaterally, but its resolution will likely take at least several years of contentious negotiations. India's role in international finance is small but growing, including the rise of the rupee toward becoming an international currency, and India's participation within the international financial system will almost certainly grow substantially over the next several years.

The backdrop to U.S. financial relations with China and India is the unfortunate fact that the international financial system, as embedded within the IMF, has become largely irrelevant to current international financial relationships. This systemic quandary, moreover, is broader than finance, extending to the international systems for trade, investment, and official development assistance. The Bretton Woods international economic system was created in the 1940s, consisting of the IMF for international finance, the GATT—currently the WTO—for international trade and some aspects of investment, and the World Bank[128]—later including four regional multilateral development banks—for official development assistance. This system has been an extraordinary success on almost all major counts, but now, largely because of its success, its modalities have become obsolete and unresponsive to current international economic relationships. A basic restructuring of the Bretton Woods system is needed, and how and why this should be done is addressed, in multilateral context, in the annexes to this and the following chapter.

The presentation in this chapter concentrates on the decisive roles of the United States and China, and to a lesser but growing extent India, to respond, over the next several years, to the serious and growing international financial imbalances and the lack of agreed systemic

[128] The official name is the International Bank for Reconstruction and Development, but the term "World Bank" is normally used for convenience, even though it designates only one part of the full institution.

norms for dealing with them. Chapter 9 then takes on the same task for international trade, investment, and official development assistance.

Currency Manipulation: IMF and WTO Obligations

As explained in the annex, the principal operational role of the IMF during its first 50 years was to provide loans to members to maintain exchange rates above market-determined levels during periods of financial stress and adjustment, but this function is now nearly obsolete. Over the past six years, the preponderant exchange rate policy has been a managed floating rate, with a strong tendency, especially in Asia, and most importantly by China, to maintain rates below a market-based level through protracted, large-scale purchases of foreign exchange by the central bank. This policy of central bank purchases leads to the currency manipulation issue, by providing the manipulating nation an unfair competitive advantage in trade through the resulting lower export and higher import prices. When a country, such as China, already has a large, sustained trade surplus, it amounts to pure mercantilism, whereby a large trade surplus becomes the national policy objective. This policy is also referred to as "beggar-thy-neighbor," whereby additional job creation at home from the trade surplus results in corresponding job losses in trade deficit trading partners. In the context of current U.S.-Chinese trade, moreover, such gains and losses in jobs are heavily concentrated in advanced technology industries, or, in other words, are an integral component of overall Chinese economic strategy to develop indigenous technological innovation and related export growth by advanced technology industries.

Both the IMF and the WTO prohibit such currency manipulation, but with a dismal record of surveillance and compliance.[129] IMF Article IV, Section 1, states that members should "avoid manipulating exchange rates . . . in order . . . to gain an unfair competitive advantage over other members." Section 3 elaborates on "the right of members to have exchange arrangements of their choice consistent with the purposes of the Fund and the objectives under Section 1 of this Article." In other

[129] For an in-depth account of IMF dismal performance, see Michael Mussa, "IMF Surveillance over China's Exchange Rate Policy," presented in draft form at a Conference on China's Exchange Rate Policy at the Peterson Institute for International Economics, October 19, 2007. Mussa characterizes IMF surveillance performance as a "catastrophic failure," and calls its application to China's exchange rate policy as "misfeasance, malfeasance, and nonfeasance" on the part of the "Managing Director and more generally by the IMF," p. 4.

words, member exchange rate policies, whether fixed, floating, or something in between, must be implemented in a way that does not entail currency manipulation, as proscribed under Section 1.

This leads to the definition of currency manipulation, which is contained in the IMF surveillance procedures related to Article IV, most explicitly in characterizing currency manipulation as "protracted large scale intervention in one direction in the exchange market," with "one direction" obviously meaning central bank purchases of foreign currencies, since this is the way to maintain an undervalued currency so as to gain an unfair competitive advantage.

This direct linkage of currency manipulation to central bank intervention in currency markets is both logical and critical. It is logical because central bank intervention is the predominant policy instrument targeted directly on influencing the exchange rate. It is critical because it totally decouples the strictures against currency manipulation from the endless theoretical debate among economists about what is an "equilibrium exchange rate." Even if economists could agree on what an equilibrium rate is, the rate would be elusive because in the current highly dynamic, globalized economy, equilibrium rates, however defined, constantly change. IMF Article IV, in contrast, focuses its surveillance on member government actions, and to decisions to intervene in currency markets in particular. This pragmatic definition of currency manipulation is reinforced by official statements, including by the United States and China, that the objective is to move to market-determined exchange rates, which can and do change over time, rather than to try to establish equilibrium rates.

Article XV of the GATT, now incorporated within the WTO, deals with "Exchange Arrangements" and stipulates that members should not take exchange rate actions that "frustrate the intent of the provisions of this Agreement." The intent of the Agreement, as stated in the Preamble, centers on the objective of "entering into reciprocal and mutually advantageous arrangements directed to the substantial reduction of tariffs and other barriers to trade." Clearly, "exchange rate manipulation to gain an unfair competitive advantage," as defined by IMF Article IV, meets the "frustrate the intent" test, because its impact on prices offsets the increased market access for exports resulting from the reduction of trade barriers. Moreover, GATT Article XV also provides for full consultation with the IMF, including the stipulation that members "should accept all findings of statistical fact presented by the Fund relating to foreign exchanges." Thus there is a direct link between IMF proscribed currency manipulation and WTO obligations for exchange rate policy under GATT Article XV.

The IMF-WTO linkage is important in terms of policy recourse for countries suffering the adverse trade effects of currency manipulation.

Even if China were found to be in violation of its IMF Article IV obligations, no specific penalties would be available if China were to continue currency manipulation. The most the IMF could do would be to make China ineligible for IMF loans, which is frivolous in view of Chinese foreign exchange holdings approaching two trillion dollars. A counterpart finding of violation of GATT Article XV, however, opens the way to WTO dispute settlement procedures, with ultimate recourse to trade sanctions against China if currency manipulation continues.

There is finally the question of "intent." Does China engage in protracted, large-scale purchases with the intent of gaining an unfair competitive advantage? This is not an issue with respect to GATT Article XV, where all that matters is the resulting impact in frustrating the mutual advantages from trade liberalization. For IMF Article IV, intent can be questioned, but an affirmative decision that the Chinese intent in central bank purchases is currency manipulation can be made without difficulty or doubt, if there is a political willingness to do so by IMF management and those members facing the adverse effects. Obviously, China and others will not officially state that they are manipulating their currencies to gain an unfair competitive advantage. But they are saying the same thing when they state, as they do, the need to keep the exchange rate below a market-based level in order to maintain satisfactory economic growth and job creation. In other words, a large trade surplus is the mercantilist policy objective, and when the trade surplus, in the case of China, soars from $100 billion in 2005 to $178 billion in 2006 to $262 billion in 2007, this is beggar-thy-neighbor trade policy on a grand scale. In these circumstances, there can be no question of the intent for maintaining an exchange rate far below a market-determined level.

Moreover, China and others need to be asked explicitly by the IMF staff, in view of the finding of currency manipulation or misalignment, what the purpose is of their protracted large-scale central bank purchases. There are legitimate reasons why members might conduct such operations. One is to replenish inadequate reserves, as South Korea and a few others did in the years immediately following the financial crises of the late 1990s. Another reason could apply to a member in sustained large current account deficit, whereby central bank purchases could be justified to help reduce the deficit. Oil exporters could also justify large central bank purchases during periods of high oil prices to avoid wide swings in their exchange rates, while in any event almost none of them would gain a significant competitive advantage in other sectors of trade because they have only small non-oil exports. But the excessive foreign exchange holdings and sustained large current account surplus of China and several others rule out the first two reasons, while the third is inapplicable. The only remaining plausible reason for protracted large-scale central bank purchases is thus to gain an unfair competitive advantage in trade, as proscribed by IMF and WTO obligations.

And yet the United States, as explained below, consistently states that China does not manipulate its currency in violation of IMF Article IV, and is silent with respect to GATT Article XV. But before assessing the policy response to date, a closer look at the unprecedented degree of recent Chinese and other currency manipulation is necessary.

Chinese and Other Asian Currency Manipulation

The test of currency manipulation, as defined by IMF Article IV, centers on the two adjectives, "sustained" and "large scale," as they relate to central bank purchases of foreign exchange. There is no question that these tests have been met by China and some others, mostly in Asia, over the past several years. Such purchases go far beyond any precedent during the 60-year course of the IMF financial system. During the first showdown with Japan over currency manipulation in the mid-1980s, the highest level of Japanese central bank purchases was $38 billion in 1987, and even these were in the context of major exchange rate adjustments associated with the 1985 Plaza Accord, whereby the yen was revalued by 80 percent vis-à-vis the dollar from 1985 to 1987.

Figure 8-1 presents the level of Chinese central bank purchases of foreign exchange for the years 2001 through a projection for 2007, together with the current account balance and the basic balance, the latter being the current account balance plus the net flow of long-term capital. The current account is the most commonly used measure of external imbalance, recording as a surplus the net inflow of foreign exchange on current account, which would have a corresponding upward pressure on the exchange rate. The basic balance, however, is a more complete, or "basic," measure of external balance, since a net inflow of long-term capital puts similar upward pressure on the exchange rate. Capital inflow is particularly important for newly industrialized economies at the takeoff stage, such as China since the mid-1990s and India more recently, because at this early stage the net inflow of growth-generating investment can be larger than the current account balance. The net inflow of capital can, indeed, offset a trade deficit from high growth in imports of capital goods related to export-oriented investment during the early years of rapid industrial modernization. This was, in fact, the case for China until about 2000 and is the current situation in India.

Figure 8-1
Chinese Central Bank Purchases of Foreign Exchange, Current Account Balance, and Basic Balance

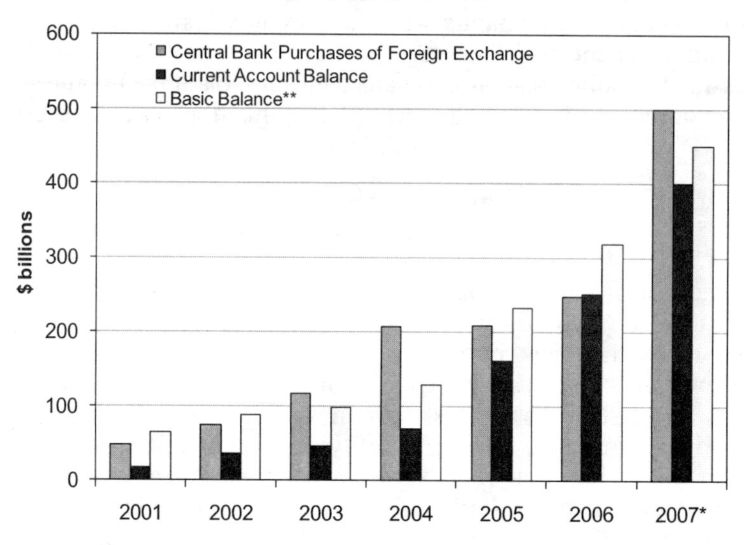

*Estimated.
**Estimated, based on current account plus estimated FDI net inflow.
Sources: IMF, *International Financial Statistics*, and *World Economic Outlook Database*, April 2007

In this context, Chinese central bank purchases, beginning in 2001, as shown in the figure, are far beyond anything that happened before. Purchases of $47 billion in 2001 were already the second highest on record after Japanese purchases of $69 billion in 2000, and then Chinese purchases soared to $117 billion in 2003, $209 billion in 2005, and an estimated $500 billion in 2007. If this does not meet the test of "protracted" and "large scale," Article IV, Section 1 should be deleted from the IMF charter. It is also noteworthy that Chinese central bank purchases track more closely with the rising basic balance surplus than with the current account surplus, although this distinction has become less important since 2005, with the current account surplus rising to become dominant within the basic balance surplus.

The Chinese Linchpin Effect

Other central banks, particularly in Asia, have also been engaged in protracted, large-scale purchases of foreign exchange and therefore have become possible currency manipulators in violation of IMF

Article IV obligations, as shown in Table 8-1. Over the four-year period 2002-2006, Japan was the second largest purchaser after China, at $463 billion, followed by India at $147 billion, South Korea at $130 billion, Taiwan at $104 billion, Malaysia at $66 billion, and Singapore at $62 billion. For all the listed economies, total central bank purchases were $2.1 trillion, and they likely rose to almost $3 trillion by the end of 2007. As a result, total reserve holdings reached $3.5 trillion by June 2007 and well over $4 trillion by year end.

Table 8-1
Central Bank Purchases, Reserve Levels, and Current Account Balance: Selected Asian Economies
($ billions)

Country	Central Bank Purchases 2002-2006	Reserve Levels June 2007	Current Account Balance 2004	2005	2006
China	1,047	1,333	69	161	250
Japan	463	914	172	166	171
India	147	214	-5	-11	-10
South Korea	130	51	28	15	6
Taiwan	104	266	18	16	25
Malaysia	66	98	15	20	26
Singapore	62	144	26	33	33
Thailand	35	73	28	-8	3
Hong Kong	24	136	16	20	20
Indonesia	20	51	2	1	10
Philippines	13	26	2	2	5
Total	2,111	3,506	302	415	539

Sources: IMF, *International Financial Statistics*, June 2007; *The Economist*, Economic and Financial Indicators, June 2, 2007; Central Bank of the Republic of China (Taiwan); and *Wall Street Journal*, July 13, 2007.

The basic test as to whether currency manipulation is the only plausible reason for such large-scale purchases is whether there is already an adequate level of reserves and a sustained large current account surplus as well. Japan, Taiwan, Malaysia, and Singapore clearly meet this test, and can thus be considered presumed currency manipulators along with China. A reasonable first-step policy response would be to press these governments to announce the cessation of central bank purchases as long as the current account remains in substantial surplus. This policy response is constrained, however, in view of the trade relationship between China and these other Asian economies, a relationship referred to as the Chinese linchpin effect. As long as China maintains its currency so far below a market-oriented level, and is the principal competitor of other regional exporters, the others will resist allowing their currencies to rise and have their exports undercut by lower priced Chinese exports. They say as much repeatedly, and the expectation is that as China finally revalues its currency, others will revalue, to a lesser extent, as well. This linkage has important implications for the large and growing U.S., EU, and other trade deficits with Asia. In 2006, 70 percent of the U.S. trade deficit in manufactures was with the Asian economies listed in Table 8-1, 40 percent with China, and 30 percent with the others. Chinese currency manipulation, as a result of the Chinese linchpin effect, is therefore both a large direct bilateral cause of the U.S. deficit and an indirect cause through other deficits in Asia.

The Three Trillion Dollar Unintended Consequence

An important consequence of Chinese currency manipulation since 2001 for the international economic system is the impact of the resulting huge increase in Chinese central bank foreign exchange holdings, from $200 billion in 2001 to $1.5 trillion in 2007, headed for $2 trillion in 2008 and $3 trillion in 2010. This apparently unintended consequence of Chinese exchange rate policy is into largely uncharted policy waters for both the international financial and trading systems, while the IMF and the WTO, as explained in the annexes, have not begun serious discussion of what is involved. In fact, there are two separate consequences, one affecting international finance and the other international investment. The financial consequence is the extraordinary leverage it provides the Chinese government to influence financial markets, and is discussed here. The second consequence is the effect of the Chinese government decision to invest large portions of its reserve holdings, starting with $200 billion in 2007, in foreign equity markets. This decision could have substantial political as well as commercial impact on international investment and on corporate management in the case of companies acquired by or absorbing large equity holdings of the Chinese government. This "sovereign wealth fund" issue is discussed in Chapter 9 with respect to Chinese and other sovereign lending institutions.

The financial leverage effect is very straightforward. With two to three trillion dollars of foreign exchange holdings spread among key currencies, China has great influence on and can cause substantial, potentially disruptive, volatility in currency markets, from shifting existing holdings from one currency to another or from changing the mix of currencies in its ongoing large-scale purchases. There are several reasons why China might do this. It could decide to have more balanced holdings rather than maintaining the estimated 70 percent in dollar-denominated assets. It could also decide to reduce dollar holdings relative to Asian and other currencies in anticipation of a devaluation of the dollar. Perhaps most threatening for the United States, China could sell large amounts of dollars, or simply threaten to do so, as political leverage to achieve other policy objectives, such as to block U.S. protectionist actions against Chinese exports or to reduce U.S. support for Taiwan. Such sales, or the threat of sales, could cause a sharp drop in the dollar, with adverse impact on the U.S. economy, which would be particularly troubling for an American administration during an election year.

In August 2007, Xia Bin, finance chief at the Chinese Development Research Center, commented that Chinese foreign reserves should be used as a "bargaining chip" in trade disputes with the United States, and another official at the Academy of Social Sciences elaborated on the remark by saying that China had the power to set off a dollar collapse if it chose to do so. President George W. Bush, in response, commented that "if that's the . . . position of the government, it would be foolhardy for them to do this."[130] The Chinese government did not make its position known, but presumably someone in the government had approved these statements beforehand.

Major dollar sales would be foolhardy at this time because the United States is a very large market for Chinese exports, which benefit from a strong dollar. This dependency is in the process of change, however, for two reasons. The first reason is that the share of Chinese exports going to the U.S. market is in steady decline, as shown in Table 8-2. The U.S. share, adjusted to include Chinese exports to Hong Kong reexported to the United States and elsewhere, declined from 27 percent in 2000 to 24 percent in 2007, while the EU share rose from 20 to 24 percent, and the share to regions other than Asia, including Africa, Latin America, and other Europe, rose from 10 to 14 percent. On this trend, the U.S. share would drop to 22 percent in 2010 and be less than 20 percent within ten years. The second reason is that as China adjusts to more domestically oriented growth, which is inevitable no matter how painful, Chinese overall growth will be even less dependent on exports to the United States.

The outlook, therefore, is for Chinese growth to become substantially less dependent on exports to the United States over the coming ten years, and this will increase the feasibility of China's shifting or threatening to shift central bank foreign exchange holdings away from the dollar, for any of the reasons stated above, including political leverage for other than financial objectives.

---

[130] See www.telegraph.co.uk and www.breitbart. com/article.php, both as of August 8, 2007.

Table 8-2
Chinese Merchandise Exports by Destination Adjusted for Hong Kong Reexports
(Percentage)

	2000	2004	2007*	2010**
United States	27	26	24	22
EU	20	22	24	26
Asia	43	42	38	34
Other	10	10	14	18

*Estimated, based on January-October.
**Projected, based on growth rate during 2004-2007.
Sources: China's Customs Statistics (Monthly Exports and Imports); and Hong Kong Census and Statistics Department.

A Rising Rupee Comes of Age

India is rising to become a major financial power, as its international trade and investment grow 25 to 35 percent per year and as the rupee becomes convertible on capital account and an internationally traded currency. This raises questions about Indian participation in the international financial system, which has been minimal since India freed itself from IMF loan dependency in 1991. The immediate issue for the next several years relates to currency manipulation, and how Indian central bank intervention, in policy terms, will become linked to what happens in China and elsewhere in East Asia. The longer term issue is how India will participate in a restructured financial relationship among the principal currency nations, which is left for the annex.

The Indian role in international finance over the next several years depends on whether India follows the Chinese path of currency manipulation, holding the rupee well below a market-oriented rate while the central bank makes protracted large-scale purchases of foreign exchange, or whether it pursues a more balanced, less export-oriented growth strategy, with a progressively rising rupee exchange rate. It is not clear which path India will follow, and little discussion of the subject within the Indian government has been reported. Large central bank foreign exchange purchases, from 2004 to 2007, brought reserve holdings to $300 billion by the end of 2007. The government, however, did allow the rupee to float up by 12 percent during 2007, in order to restrain inflation. This revaluation received the usual complaints from export industry, while having the positive effects of lower import prices and a reduced foreign debt, in rupees, for Indian companies.

The analytic setting outlined in Chapter 5 is that India is in the high growth, takeoff stage of industrial modernization, with a large net inflow of long-term investment, both FDI and portfolio, while the current account is in relatively small deficit. This dynamic, high growth course of external accounts has been clouded by the inadequacy and time lags in Indian statistics. A large merchandise trade deficit, for example, gets prominent media coverage, while statistics on the largely offsetting trade surplus in business services lag by one to two years and are likely understated. Figures on the current account and long-term capital flow face similar time lags. The best estimate of the course ahead for 2008 and 2009, given these statistical constraints, is for a continued large and possibly growing net inflow of long-term capital together with a shift of current account from relatively small deficit into significant surplus, as the recent surge of investment in the manufacturing and other export-oriented sectors translates into rapid export growth. This, in turn, will result in a large and growing net inflow of foreign currencies, on basic balance account, and raises the question of what the Indian central bank will do: will it make protracted large-scale purchases, that is, currency manipulation, or will it make restrained purchases while allowing the rupee to rise?

The actual scenario for India will be influenced greatly by what China does on the revaluation path toward more balanced growth. India is another participant in the China linchpin effect. Indian exports increasingly compete with Chinese exports in third country markets, while FDI decisions are influenced by the relative salaries of engineers and other costs in China and India. Chinese exports are also flooding the Indian market, up 63 percent in 2006 and 66 percent in 2007. Consequently, India will be reluctant to allow the rupee to appreciate faster than the yuan, as happened during 2007. When China finally begins along a more substantial revaluation path, however, India, like others in East Asia, should become disposed to allow at least lesser revaluations of its currency.

In this uncertain context and with growing pressures to confront the currency manipulation issue, India will likely seek to play a low-key role in IMF deliberations, while attempting to maintain its freedom of action at this still early stage of sustained high national growth. It is not too early, however, for India to engage in serious discussion within the government, and with major trading partners, most importantly the United States and China, as to how India will or should participate more actively in the international financial system. Moreover, on the basis of current and projected high growth of Indian GDP, FDI, and exports, there should be no doubt that within 5 to 10 years India will become a major financial power and will have to adopt policies in keeping with this new role.

The Policy Response of Denial and Unconcern

The policy response through mid-2007 by the United States, other major trading nations, and the IMF to surging Asian currency manipulation can be characterized as in denial and of unconcern. There were various reasons for this nonresponse. The broadest reason was that the global economy grew at an exceptionally robust rate of 4 to 5 percent from 2003 through 2007, driven largely by 10 to 11 percent growth in China together with the willingness of the Chinese central bank to make large-scale purchases of foreign exchange to finance high growth elsewhere. Currency manipulation can have short-term benefits for growth, however unbalanced and trade-distorting the longer term effects.

The decisive role in the policy response of denial was played by the United States. The U.S. Secretary of the Treasury issued a statement every six months denying that China or any other nation was manipulating its currency in violation of IMF Article IV. The EU, meanwhile, was essentially unconcerned, viewing the issue as a U.S.-China bilateral problem related to the growing U.S. trade deficit. Japan avoided the issue because it, too, was being accused of currency manipulation by some in the United States. IMF management, mindful of the categorical denial of currency manipulation by its largest and most directly affected member, was careful never to mention the term. And China, in response to criticism from the U.S. Congress, simply stated that exchange rate policy was a matter of national sovereignty, thus explicitly repudiating its obligations under IMF Article IV and GATT Article XV.

This policy of nonresponse and lack of concern began to change in mid-2007, as explained below. First, however, in view of the decisive U.S. role, some elaboration of its recent U.S. policy course is in order. Current U.S. policy derives from the Omnibus Trade and Competitiveness Act of 1988. Congress had been displeased over the weak response of the Reagan Administration to East Asian currency manipulation, especially by Japan, and inserted a provision in the Trade Act requiring the Secretary of the Treasury to report to the Senate Banking Committee every six months as to whether any nation was manipulating its currency to gain an unfair competitive advantage. The link to IMF Article IV was not explicit in the Act, but had been absolutely clear during discussion of the draft legislation. During the initial years through 1994, the Treasury cited Taiwan, South Korea, and China as currency manipulators, but since then no nation has been cited. The currency manipulation issue, in any event, receded as the U.S. trade deficit declined sharply following the dollar devaluation of

1985-1987, while the financial crises in East Asia and elsewhere in the late 1990s related to over- rather than under-valued currencies.

Beginning in 1999, the U.S. trade deficit began to surge again, with China out front, as explained in Chapter 3. Pressures for a strong U.S. reaction quickly developed in Congress, including unanimous resolutions in both houses stating that China and others were manipulating their currencies in violation of IMF obligations and that the president should take more forceful actions to stop it. Broadly based private sector complaints grew in parallel.[131] The Treasury response, however, was to continue its semiannual denial of currency manipulation by any nation, while engaging in bilateral dialogue with China to seek restraint in the burgeoning bilateral trade deficit. U.S. denial of currency manipulation by China was explained by the lack of stated intent by China, which had little credibility with the Congress and other critics.

There are various possible reasons why the Secretary of the Treasury continues to deny currency manipulation by China and others. There is the official judgment that more can be achieved through private, bilateral talks than through multilateral dispute procedures. Influential conservative opinion molders, such as the *Wall Street Journal* and the Cato Institute, consider the trade deficit, no matter how large or how concentrated in advanced technology industries, a nonproblem and nothing to worry about. Some American banks and companies with production in China benefit from stable exchange rates and an undervalued yuan. Foreign policy objectives, such as Chinese collaboration to suppress nuclear weapons programs in North Korea and Iran, can be overriding, and used as a reason to avoid economic conflict with China. Perhaps most important, actions to curtail Chinese currency manipulation and the growing trade imbalance will involve difficult economic adjustment, with some transitional adverse impact, in the United States as well as in China, including, for example, higher interest rates and inflation in the United States to the extent that China and other currency manipulators reduce purchases of U.S. Treasurys and import prices rise. The problem with this reasoning is that the longer the adjustment is postponed, the greater the ultimate adverse impact is likely to be.

For these various reasons, the United States has proceeded with extensive dialogue with China for several years while achieving no significant results. China agrees that the long-term objective is for the

---

[131] The Coalition for a Sound Dollar was formed in 2002, chaired by the National Association of Manufacturers (NAM) and consisting of over 90 private sector industry, agriculture, and labor organizations, to protest currency manipulation by China, Japan, Taiwan, and South Korea, with documentation provided in its Monthly Asia Currency Manipulation Monitor.

yuan rate to become market-oriented, with the clear implication of substantial revaluation in view of the huge and growing Chinese current account surplus. But the yuan rate since 2000, on a weighted basis with principal trading partners, has actually declined through its close linkage to the dollar.

The failure of U.S. policy to reduce currency manipulation, and the consequent rise in the U.S. trade deficit, can be attributed, in large part, to especially wrong-headed positions on two issues. Fortunately, these positions began to change during the course of 2007.

The first issue is the habit of the U.S. administration to place the blame for protectionism on members of Congress and others who call for import sanctions against China and other currency manipulators even as a last resort, as provided for in the WTO, if they do not cease manipulating their currencies to gain an unfair competitive advantage. As explained above, the currency manipulators are the protectionists, indeed unprecedented protectionists, and they should be officially designated as such, rather than those who threaten retaliation as a last resort against such protectionism. A broader understanding of where the protectionism lies has become apparent, as the Chinese trade surplus surges not only with the United States but also with Europe and other regions. It will nevertheless be difficult to redirect the protectionist argument after the U.S. government itself held the American side to be largely at fault for so long.

During 2007, three draft bills were introduced in the Congress that would pressure the president to act against Chinese currency manipulation. All three are directed at IMF and WTO provisions and do not entail unilateral trade sanctions by the United States.[132] In keeping with the recommended policy response that follows, the president should state clearly that these are not protectionist bills but responses to protectionist currency manipulation by others. Such a statement would result in broad bipartisan if not unanimous congressional approval of whichever bill reaches the President's desk in 2008.

The second issue has been the almost exclusively unilateral U.S. approach for dealing with China and other currency manipulators, and the almost total avoidance of the multilateral institutions, the IMF and the WTO. A crisis showdown within these institutions would not have been a wise policy, but pursuing the currency manipulation issue and the resulting adverse impact on trade within the consultative and dispute resolution procedures of these institutions, through concerted

[132] See Gary Hufbauer and Claire Brunel, "The US Congress and the Chinese Yuan" (Presented at a Conference on China's Exchange Rate Policies at the Peterson Institute for International Economics, Washington, October 19, 2007). The three bills are H.R. 2942 Ryan-Hunter, S. 1607 Schumer-Grassley-Graham-Baucus, and S. 1677 Dodd-Shelby.

initiatives to the extent possible, while intensifying outside bilateral and small group discussion, would have been a much stronger policy framework for achieving results. U.S. unilateralism has also been in decline during 2007, as the adverse trade impact from currency manipulation deepens and broadens in geographic scope, particularly in Europe, and as the United States has come to realize that the IMF can provide institutional support for dealing with the problem. One result was the IMF mid-year initiative to strengthen surveillance procedures related to misaligned currencies, and country assessments of currency misalignment by IMF staff are under way.

This and other initial steps to deal with exchange rate policy and currency misalignment came together at the October 2007 annual IMF ministerial meeting, although principally in the G-7 and informal talks, while the official deliberations unfortunately focused on the short-term global economic outlook and the need to realign IMF voting rights. A far more forceful and concerted multilateral action program for the next several years is necessary in order to avoid even larger and more threatening imbalances in trade and official financial holdings, and to begin reducing them.

A Recommended U.S. Policy Course Ahead

The United States needs a forceful and comprehensive policy response to deal with the immediate problem of greatly misaligned currencies, centered on the Chinese yuan, and to begin a process of restructuring the international financial system in accord with the new realities of international financial policies. The presentation here addresses the immediate problem of misaligned currencies over the coming one to three years, while longer term systemic restructuring is left for the annex.

The U.S. policy response should begin with a clear statement of the problem and the policy objective by the Secretary of the Treasury, along the following lines:

> Currencies have drifted far out of line over the past five years, the Chinese yuan most of all, principally as a result of protracted, large-scale central bank purchases of foreign exchange that result in exchange rates being held well below market-based levels. As a consequence, the United States is suffering substantial adverse trade impact, most importantly for advanced technology manufacturing and related business services industries. These industries account for 90 percent of civilian R&D and patent applications, of vital interest to long-

> standing U.S. leadership in technological innovation. Such large-scale central bank purchases constitute a mercantilist policy, proscribed as currency manipulation within the IMF, and must be phased out. All nations can benefit greatly from the rapid expansion under way in international trade and investment, but such expansion must be fair and balanced and certainly not distorted by such mercantilist exchange rate policies. The United States will work closely with major trading partners who also suffer the adverse effects of currency manipulation to achieve decisive results over the coming three years. The critical benchmark for results will be a substantial reduction and ultimate elimination of such central bank purchases, particularly by trading partners who also maintain a large trade surplus in directly competing manufactures and business services.

From this broad statement of the problem and policy objective, the action program should be on three mutually reinforcing tracks: the multilateral IMF and WTO track, the G-7 extended group track, and the bilateral track.

The Multilateral Track

The United States should develop a joint approach for dealing with the currency misalignment problem within the IMF and the WTO. The EU and Canada should be fully in accord because they are equally vulnerable to rising trade deficits from currency manipulation, and even more so than the United States to the extent that their floating rate currencies rise relative to the dollar, partly as a result of China and others shifting central bank purchases away from the dollar and more heavily to the euro and the Canadian dollar. Since mid-2007, European leaders have been more outspoken in condemning Asian currency manipulation, although it is unclear how EU member governments, the European Central Bank, and the EU Commission would represent European interests in the IMF and the WTO related to exchange rate policy. Other trading nations, such as South Korea and Mexico, who have resisted large-scale central bank purchases while experiencing surging imports from China, are also likely to support actions to reduce such purchases.

The IMF objective should be to place currency misalignment and central bank intervention in currency markets at center stage within the international financial system, which is where it belongs in the new order of managed floating rates. Top priority should be given to elaborating the strictures against currency manipulation contained in

Article IV, through a strengthening of the surveillance procedures and specification as to what happens when currency manipulation is determined to be occurring. The prima facie evidence for currency manipulation of protracted and large-scale purchases should be reaffirmed. The issue of intent should also be clarified in terms of specific acceptable reasons for large-scale purchases, which would exclude the objective of creating export-oriented jobs at the expense of jobs in importing countries. Members with sustained large current account surpluses and a reasonably high level of reserves should be presumed to be currency manipulators. As for phase-out of the current misalignment, a three-year transition should be adopted, with a declining level of central bank purchases as the decisive year-to-year benchmark.

The WTO objective should be to engage in high-level discussion of the impact of currency misalignment on WTO market access obligations and how GATT Article XV should be applied under the new circumstances of managed floating rates and related currency manipulation. There is precedent for discussing a more direct linkage between the trading and financial systems, including an EU initiative at the outset of the Uruguay Round of trade negotiations, as explained in the annex, and there is now a much stronger analytic case for doing so. The relationship of currency manipulation to the WTO export subsidy agreement could also be discussed. A specific objective for the United States and its collaborators would be to state the intent to file a complaint about currency manipulation or misalignment within the Article XV dispute procedures, with a reasonable deadline, if substantial realignment of currencies and stronger strictures against central bank purchases have not been achieved within the IMF. In particular, if currency misalignment is assessed by the IMF, under the new surveillance procedures, this would be used as the basis for pursuing a GATT Article XV complaint within the WTO.

These various actions within the multilateral institutions to curb and ultimately eliminate the adverse trade effects of currency manipulation should not be limited to China, but should be pursued against several others of the most evident currency manipulators as well. Based on the figures presented in Table 8-1, this might include Japan, Taiwan, Malaysia, and Singapore. The distinct Japanese situation is described below. Taiwan is unique, not only in being one of the longest-standing currency manipulators, but also in being a member of the WTO but not of the IMF. This would create a procedural challenge if Taiwan were included in a WTO Article XV complaint procedure, because the IMF is supposed to do the statistical assessment of the complaint, but might be constrained from doing so for a nonmember.

The G-7 Extended Track

The G-7 is the grouping of principal international currency nations—for the dollar, the euro, the yen, the pound sterling, and the Canadian dollar—and is also the most practical and flexible forum for developing and implementing a strategy to phase out currency manipulation. There should be broad consensus among six of the seven—the United States, Germany, France, the United Kingdom, Italy, and Canada—inasmuch as they all have basically free-floating currencies that are vulnerable to the adverse trade effects of widespread currency manipulation in Asia.

The seventh member, Japan, is distinct, in that although it is a long-standing large-scale purchaser of foreign exchange and is generally viewed as having a substantially undervalued currency, its central bank has not made large purchases since 2005. The possible reasons for this were discussed earlier, including the yen carry trade and the credible threat of unlimited central bank purchases if the yen should rise significantly. It should nevertheless be in the Japanese interest to work with the other six to resolve the major currency misalignments caused by recent Chinese and other purchases, especially as related to its new number one trading partner, China. The basis for G-7 solidarity including Japan would be a Japanese commitment to continue to refrain from central bank purchases as long as the Japanese current account remained in large surplus.

The G-7 framework would be extended to include other major currency nations, as it has been in recent years, and these others would be China, India, South Korea, Mexico, Brazil, Russia, and perhaps a few more. Together with the seven principal members, this would represent the preponderant power structure for international finance, and would also be a practical, limited size forum for dealing with both the immediate currency misalignment problem and the longer term objective of restructuring the financial system.

The positions of the "extended" members on the currency misalignment/manipulation issue would vary. Some, such as South Korea and Mexico, as explained earlier, would likely align themselves with the G-7. China, obviously, would be on the defensive and resist specific commitments, including date-specific benchmarks. It is less clear how India, Brazil, and Russia would react, and among these, India would be the most important in view of its rapid rise toward becoming an international currency nation. In any event, the relationships of these countries with the United States and other members of the G-7 would be developed largely through bilateral discussions.

The Bilateral Track

U.S. bilateral discussions would be of critical importance, particularly for developing a concerted strategy within the G-7 and for dealing with China as key to resolving the currency manipulation issue. Other bilateral dialogue would also be an essential part of the overall strategy. The two most important bilateral relationships would be with China and Japan. The least clearly defined at this stage would be with India.

China.—The United States and China already have the structured Strategic Economic Dialogue, which would become more directly engaged in dealing with the exchange rate issue. The mutually agreed objective of moving to market-oriented exchange rates should be reaffirmed and linked to a timetable for results. The indicators for results should be a reduction in Chinese central bank purchases and current account surpluses, which are estimated to have soared to $500 billion and $400 billion, respectively, in 2007. The principal means for doing this would be a progressive revaluation of the yuan, starting with at least 20 percent in the first year. If this did not achieve significant results for the indicators, further revaluations would take place in years two and three. Such a three-year scenario would be similar to the Japanese 80 percent revaluation in 1985-1987, when the global imbalances and currency misalignments were smaller than they were in 2007. As explained in Chapter 5, such an adjustment away from excessively export-oriented growth would be very difficult for China, and the United States should offer to take supportive actions to facilitate the structural change within China.

Japan.—The issue of undervaluation of the yen despite Japan's recent cessation of central bank purchases would be a major subject of bilateral discussion. The immediate objective, as noted above, would be an officially announced cessation of central bank purchases as long as the Japanese current account remained in substantial surplus. The discussion would also address concerted economic strategies that would lead to sustained Japanese economic growth in the context of a declining trade surplus. This could most effectively be done through the parallel negotiation of a bilateral free trade and investment agreement, as discussed in Chapter 9.

India.—This should be the most interesting bilateral dialogue intellectually, with heavy focus on the progressive rise of India to become a global financial power, and on how India, as a result, should play a growing leadership role in the international financial system. The issue of restraint in purchases by the Indian central bank would emerge to the extent that the Indian current account shifts into a growing surplus, which is likely over the coming three years. As with Japan, the bilateral financial dialogue would benefit from discussion, in parallel, of a bilateral free trade and investment agreement.

The Highly Uncertain Step-by-Step Scenario Ahead

The foregoing three-track policy initiative should produce a three-year realignment of key currencies, based principally on market forces, and the restoration of greater balance in trade. This result would also set the stage for a restructuring of the international financial system, not only to maintain such market-oriented exchange rate relationships but also to deal more effectively with other emerging issues in international finance. The one to three year scenario outlined here, however, even if pursued with vigor by the G-7 and other like-minded nations, would face great uncertainties as to the outcome, as well as substantial threats of major financial turbulence, with potential adverse impact for particular economies and the global economy in general. The two greatest uncertainties which will likely be decisive for the outcome would be the responses of China and financial markets.

The Chinese response.—China may simply continue to stonewall on substantial revaluation of the yuan, perhaps reluctantly revaluing by about 10 percent, but with no intention of going further. The result of this would almost certainly be little or no impact on the upward trend in the Chinese trade surplus, particularly for politically sensitive manufactures, while central bank purchases would soar to holdings of $2 trillion by 2008 and $3 trillion by 2010. This result, in turn, would intensify protectionist pressures in the United States, Europe, and elsewhere, and, within the scenario presented here, trigger a WTO dispute procedure against China under GATT Article XV. The net result would be a continued currency misalignment/manipulation impasse into 2009, only in larger, even more threatening terms for the global economy.

The central dilemma for China is how to restructure its economy away from excessively export- or, more pointedly, mercantilist-oriented growth toward domestically oriented growth. Chapter 5 explained this dilemma in detail, with a projected painful "hard landing" ahead in economic terms, which would, in turn, increase pressures for political change within China.

There is no way to predict which way the current, highly risk-averse communist leadership in China will go on these issues over the next one to three years. There is a clear U.S.-China mutual interest in moving forward with currency realignment and related structural adjustments in both national economies, while minimizing the adverse transitional costs. Unfortunately, the vaunted ministerial level Strategic Economic Dialogue of 2006-2007 has not been up to the task of addressing these issues in serious, concrete form.

The financial market response.—This is even more difficult to predict, but a substantial initial revaluation of the yuan could trigger a volatile financial reaction that would overwhelm ongoing dialogue among governments. This risk will only grow to the extent governments are unable or unwilling to take effective action to deal with the growing currency misalignment. The most likely market reaction in these circumstances would be for markets, at some unpredictable point, to anticipate a substantial further devaluation of the dollar and shift to relatively larger holdings of assets in other currencies, principally the euro. To some extent they are already doing this, as the euro rose toward 150 to the dollar in 2007. If the rate were to rise to 160 or 170, there would be a strong political backlash in Europe, which could interact with a corresponding growth of protectionist pressures in the United States from the ever-larger trade deficits with China and other Asians. Large central bank purchases by the Europeans to prevent further upward movement of the euro and other European currencies, purchases that would likely have to run in the hundreds of billions of dollars, and a likely U.S. protectionist counteraction, could unleash still larger and more destabilizing movements in private financial markets.

The financial market reaction would be highly uncertain even under the three-year scenario outlined above that would credibly lead to a major reduction in the currency misalignment. A first-year revaluation of the yuan by 20 percent with the prospect of further, perhaps larger revaluations in years two and three if Chinese central bank purchases did not decline could quickly lead markets to anticipate the further revaluations and shift their holdings into yuan- and other Asian currency-denominated assets, with a reduced willingness to finance the still huge U.S. current account deficit. The result could be a more rapid decline in the dollar and in the overall exchange rate realignment. This is what happened in 1985, when modest intervention by the G-7 central banks was followed by a much stronger reaction, to some extent an overreaction, in financial markets. Indeed, the financial market role in many past step-by-step official scenarios for phased exchange rate adjustment has been to greatly accelerate the adjustment process.

In sum, the actual outcome cannot be predicted, although the directions of change are clear and governments need to gear up for a period of more active financial diplomacy ahead. Strong and concerted leadership by the G-7, still the central grouping of global market-oriented currencies, will be essential. China, likewise, needs to take greater responsibility for the trade and financial impact its undervalued currency is having. And India, the incipient global currency nation, will be greatly affected over time by the results, must soon play by the rules of the international financial game, and therefore has an interest in helping to shape the new order of international financial relationships.

Annex 8

The International Financial System in Need of Reform

The IMF has lost touch with the current realities of exchange rate relationships and is thus in default as the center of the international financial system. A fundamental reform is necessary to restore its management role. This sweeping statement requires an explanation of how the IMF got from its beginnings to where it is today. This is followed by an annotated agenda for the proposed reform.

A Brief History of the IMF System

The IMF was created at Bretton Woods in 1944 and was officially launched in 1947. It began as a dollar-based system, with other exchange rates fixed but adjustable to the dollar, and with the dollar linked to gold. How and when other rates would be adjusted was not fully defined. Members could not adjust the par value of their currencies by more than 10 percent without the permission of the Fund, but implementation of this requirement was weak to nonexistent.

Within this framework, the central role of the IMF during its first 25 years was to provide relatively short-term loans to members trying to maintain the dollar rate during periods of economic adjustment and financial stress. IMF loans, in effect, helped members to keep their exchange rates higher than market forces would have dictated. There was little concern about exchange rates to the dollar being too low because, until the 1960s, there was an overarching "dollar shortage," and the United States maintained a seemingly permanent trade surplus.

During these initial years, the large majority of IMF loans were to industrialized countries, including to the United States, and the results were highly successful. This was the post-war period when the industrialized grouping moved to currency convertibility and greatly lower trade and investment barriers, while IMF loans helped stabilize financial markets and exchange rates during the transition. Developing countries were far less engaged because, for the most part, their currencies were not convertible and trade protection remained very high.

Then in the 1970s two major changes took place. First, U.S. trade had gone into deficit and the dollar shortage ended, largely because some European and other trading partners maintained exchange rates below market-based levels. As a result, dollar holdings abroad rose and were converted into gold through purchases from the U.S. Treasury. The United States responded in August 1971 when U.S. Secretary of the Treasury John Connally and his young Under Secretary for Monetary Affairs Paul Volcker declared an end to U.S. gold convertibility, imposed a 15 percent import surcharge to help reduce the U.S. trade deficit, and demanded that others allow their currencies to rise to market-based levels. The United States did not explicitly accuse others of currency manipulation, but, in current parlance, that was what the issue was. Others claimed the import surcharge was in violation of GATT obligations, but agreement was soon reached, European rates were adjusted upward, and the U.S. import surcharge was lifted.[133]

This confrontation led to a system of highly flexible and floating rates among the industrialized countries, and declining recourse to IMF loans. The last major IMF loans to this grouping were in 1976, to the United Kingdom and Italy. In effect, IMF members accounting for two-thirds of international finance had "graduated" from dependence on IMF loans. Currency convertibility together with open and rapid growth in trade and international investment had engendered relatively stable financial markets, served best by market-oriented, lightly managed floating exchange rates. The IMF lending role, as a result, was greatly diminished, which should have been celebrated as a major success for the overall Bretton Woods system.

The second major change in the 1970s was the emergence of "newly industrialized" economies, first in East Asia and then spreading to Latin America, South Asia, the former Soviet Bloc, and elsewhere. They pursued a similar strategy as the industrialized grouping had done earlier, of progressive currency convertibility, trade liberalization, and openness to foreign investment. In the early stages, financial markets were similarly weak and volatile, and IMF loans proliferated to maintain exchange rate stability during periods of market-oriented economic and financial reform. Over time, however, the IMF loans came to be used more and more to postpone difficult reforms, and the IMF, in response, tightened policy conditionality to pressure governments to move often controversial reforms forward. The result was that the ultimate

---

[133] As a historical footnote, the author was a member of the U.S. delegation that was sent immediately to Geneva to defend the import surcharge before the GATT. In fact, the United States was in violation, since the GATT permits temporary import quotas but not tariffs in response to a trade deficit. The U.S. delegation, however, treated this distinction as of small consequence, while emphasizing that the problem had to do with exchange rates. Thus, one of the first official exchanges on what has become today the central problem facing the international financial system of currency manipulation took place in an obscure GATT working party.

adjustment was often too late and more costly. This whole process came to a head in the financial crises of the late 1990s, including in Thailand, Indonesia, South Korea, Brazil, Argentina, Turkey, and Russia. In each case, a higher than market-oriented exchange rate was prolonged through IMF and other official borrowing, followed by financial crisis, with IMF and other debt having to be repaid more painfully with a devalued currency.

By 2000, the newly industrialized grouping had learned its lesson and began to phase out IMF loans. They had also, for the most part, achieved more balanced growth, open to trade and investment, with deeper and more stable financial markets. They, too, had successfully graduated from the IMF loan stage of industrial development. Ninety percent or more of international financial transactions were now by IMF loan graduates, an even more striking success story.

In parallel with this mission accomplished experience for the IMF loan program, the issue of currencies that had been deliberately undervalued to give a competitive trade advantage, the issue that had triggered the U.S. reaction in August 1971, became much more important. The second major confrontation over this issue came in the mid-1980s, when Japan, Taiwan, and some Europeans engaged in large-scale central bank purchases together with rising trade surpluses with the United States. The problem was resolved through the 1985 Plaza Accord, whereby coordinated central bank intervention by the key currency nations led to a 40 percent decline in the dollar and increases vis-à-vis the dollar of 80 percent for the yen and about 50 percent for European currencies.[134] These large and to some extent excessive swings in exchange rates were caused principally by financial markets piling on to what were only modest official interventions, an important interacting relationship relevant to the current situation of misaligned exchange rates. Another significant observation about the experience of the 1980s was that official negotiations took place at the eponymous Plaza Hotel in New York, totally outside the IMF framework. The issue was basically currency manipulation as defined by IMF Article IV, but it was never raised within the IMF, although it did make its way into the post-Plaza U.S. Trade Act of 1988.

The next and current phase for the issue of undervalued or manipulated currencies began in the late 1990s, emanating principally from the recovery stage of the financial crises in the newly

[134] As noted earlier, percentage changes are larger for the revaluing than for the devaluing currency. For example, a 50 percent devaluation of the dollar to the yuan results in a 100 percent revaluation of the yuan to the dollar.

industrialized grouping. Undervalued exchange rates became very popular with politically powerful export industries, and proved to be more effective than IMF loans for achieving high, export-oriented growth, albeit on a beggar-thy-neighbor basis. China anticipated this strategy by devaluing its currency in 1994, which enabled it to weather the financial turbulence late in the decade, and it has essentially maintained that undervalued peg to the dollar through 2007.

A noteworthy aspect of this recent phase of the currency manipulation problem is that again, as in 1971 and in the mid-1980s, it has been taking place outside the IMF framework, at least through mid-2007. The United States has been most emphatic in dealing with the issue on an almost exclusively bilateral basis, particularly with China, with some consultation within the G-7, but virtually no serious discussion in the IMF.

This whole 60-year saga of the evolving international financial system—a great success story assuming the current currency mis-alignment problem can be resolved—raises the question of how the IMF as an institution should address the changed pattern of exchange rate relationships that has evolved since 1971, so as to adapt and remain at the center of the financial system management process. In fact, there was almost no serious discussion of this issue within the institution, while the IMF loan process continued to dominate time and attention. This essentially backward-looking posture was most evident during preparations for the 50th year anniversary of Bretton Woods in 1994. A Bretton Woods Commission of 47 distinguished financial leaders and experts, chaired by Paul Volcker, the intellectual author of the first U.S. assault on currency manipulation in August 1971, called for the "establishment of a new system . . . [because] the alternative to the new global system is to continue the present nonsystem." The problem was thus explicitly stated, but the Commission report had little to offer as to what form the new system should take except to note that the "system could possibly involve flexible exchange rate bands."[135]

Five months later, the Mexican peso crashed through the bottom of its exchange rate band with the dollar and financial markets assumed the lead role in pushing governments toward a truly new, post-dollar, predominantly managed floating rate system by decade end. This new system, as explained above, greatly reduced the IMF lending program, while bringing to center stage the issue of how members manage their highly flexible or floating rates, including the strong trend toward protracted, large-scale central bank purchases, with consequent

[135] This proposal, which raised more questions than it answered, was supported by a minority of the Commission, spearheaded by C. Fred Bergsten.

undervalued currencies. The new order of exchange rate relationships clearly raised a number of important questions, although little discussion of them took place within the IMF system.[136]

In recent years, absent the need for decisions about large IMF loans, the IMF has concentrated its attention, up through ministerial-level meetings twice a year, on such issues as forgiveness of relatively small amounts of IMF loans to least developed countries that would never have repaid the loans in any event. There has also been prolonged debate over a redistribution of voting rights and the procedure for selecting the managing director, in a situation where there is little of consequence to vote on or manage. The largest emerging financial issue within the institution is how to pay for salaries and other costs of an IMF staff that was greatly expanded during the financial crises of the late 1990s, and which had always been paid for through interest on IMF loans that had now evaporated along with the loan program. Annual meetings draw media-enthralled crowds of protesters, but there is little of substance to protest about regarding the IMF.

Finally, in 2007 the IMF began to address the growing currency misalignment issue, centered on China, in terms of initial steps toward broader institutional reform, especially concerning surveillance

---

[136] An early outside warning of the new realities was offered, but it received scant recognition. See Ernest H. Preeg, *The U.S. Trillion Dollar Debt to Foreign Central Banks* (Institute for International Finance, 1998). The essay concluded that the financial system faced "a systemic dilemma for the long-standing dollar-based . . . system." The central problem was "mercantilist-motivated official dollar purchases with U.S. acquiescence. . . . From 1992 until 1997, official dollar holdings . . . increased $139 billion for Japan and $91 billion for China, even while both countries maintained current account surpluses. . . . To the extent East Asian economic growth recovers by late 1998 and 1999, based on depreciated currencies and ever more export-oriented growth strategies, the temptation of central banks to buy dollars to avoid an excessive strengthening of their currencies will likely reemerge, perhaps with a vengeance." A particular problem was "foreign government leverage against the United States . . . China, for example, could threaten the sale of official dollar holdings as leverage for foreign policy or other objectives." The proposed solution was "a more balanced structure of foreign exchange holdings by central banks, including disciplines to limit mercantilist-motivated reserve accumulation." Ten years later, nothing has changed except that the magnitude of the problem is several times larger, including the August 2007 Chinese threat to sell dollars. The same proposed solution is elaborated in the following section of this annex. The original essay won second prize in the competition in honor of Jacques de Laroisière, and one of the jurors was Paul Volcker.

procedures related to Article IV, although explicit reference to currency manipulation remains politically incorrect. A far broader agenda for fundamental reform and renovation is needed, however, and it probably needs to begin with the formation of a specially constituted, independent grouping of forward-thinking wise persons, located outside the organizational structure of the IMF. The landmark hotel in Bretton Woods, New Hampshire, would be both a scenic and symbolic setting for launching the reform process. The agenda for such a grouping could be along the following lines.

An Agenda for IMF Reform

The starting point for IMF reform would be agreement that the new order of exchange rate relationships should be market based, with exchange rate policies designed to support such an orientation. This start presupposes that credible initial steps are taken to reduce the current misalignment of currencies. But beyond this, there is such agreement, in principle, not only among the G-7 key currency nations, but between the United States and China as well.

Within this basic understanding, the first step toward reform should be a statement of the extraordinary success of the IMF system over the past 60 years, and how this success has changed the policy framework for international financial relationships and now requires basic reform of the IMF so that it can regain its central management role. The IMF has successfully supported the rapid expansion of international trade and investment to the great benefit of the very large majority of the global population. This very success, however, has created a new situation resulting from the shift away from fixed but adjustable exchange rates that were sustained in times of financial stress by substantial foreign exchange holdings and IMF loans, and toward a new system of lightly managed floating or highly flexible rates, with a greatly reduced need for such foreign exchange holdings and IMF loans. The policy focus of the new system should consequently be on central bank intervention in currency markets as the principal policy instrument for managing exchange rates, and the corresponding IMF role should be to ensure that this is done in a concerted, mutually beneficial manner, supportive of the objective that exchange rates should be market oriented.

The specific reform agenda for IMF renovation that will respond to this new situation can be formulated in terms of three major issues where important change is vital, and six other issues of lesser concern because they require less change or are simply less important. The aggregate set of substantive changes then leads to the need for significant organizational change for the IMF as an institution.

The Three Major Changes

The three issues for major change are disciplines for central bank intervention, disciplines for the use of foreign exchange holdings, and linkage between the financial and trading systems.

Disciplines for central bank intervention.—This is the currency manipulation issue as contained in IMF Article IV. Disciplines for central bank intervention need to be clarified and made more specific. Surveillance procedures need to be strengthened for identification of violations. The point of departure for a system of lightly managed, principally market-oriented exchange rates, is that there would be relatively little need for central bank intervention. This is the current practice for a number of major currencies, including the U.S. and Canadian dollars, the euro, and the pound sterling. Existing IMF prima facie identification of currency manipulation as protracted, large scale central bank purchases should be considered presumptive evidence of such manipulation, except where overriding circumstances are presented, subject to a review process. One such circumstance that would require careful formulation is that of oil exporting nations during periods of high oil prices. The general rule would be that members in sustained current account surplus should refrain from central bank purchases, and they would be encouraged to gradually draw down excessive reserve holdings in support of greater balance in external accounts among all members. The definition of "excessive" leads to the second major issue.

Disciplines for the use of foreign exchange holdings.—This will require an even more fundamental review of necessary changes in the international financial system, because the whole purpose of official exchange holdings has now largely been overtaken by events, while holdings, in absolute terms, have skyrocketed. The original purpose for holding "adequate" or "prudent" reserves was to be able to maintain a fixed rate during periods of financial stress through sales of foreign exchange at the official rate. The World Bank benchmark for adequate reserves was 25 percent of annual imports. A lightly managed floating rate, however, would require a far lower level of reserves for smaller and less frequent interventions, and no reserves at all for a freely floating rate. The extraordinary low level of U.S. foreign exchange holdings of $45 billion is in keeping with the new situation, while at the other extreme the rising level of Chinese reserves toward $2 trillion and $3 trillion is the outstanding manifestation of the systemic dilemma facing the IMF system.

Reform of the role of official reserves in the new situation therefore needs to begin with a redefinition of the purpose of official reserve holdings, within a system of lightly managed market-based rates. The definition would clarify the need for a much lower level of reserves related to trade, although a precise percentage figure would not be necessary.

The disciplines for foreign exchange holdings, based on the new definition, and in view of the huge buildup of reserves by a number of countries in recent years, would fall into three categories. The first is transparency. The only current requirement is that IMF members report on a monthly basis the total level of foreign exchange holdings, while the composition of the reserves by currency is kept secret. Even these total figures can now be misleading and greatly understated as governments shift official foreign exchange holdings to various forms of sovereign investment funds. Monthly reporting should be extended to include a breakdown by major currency holdings, and large interventions should be reported to other members as they take place. With unprecedented large total reserves by a number of countries, substantial shifts in reserves from one currency to another, for whatever reason, can have a disruptive effect on currency markets. Other members and private sector participants in currency markets should be aware of central bank actions that could have such disruptive impact on markets.

The second category of disciplines relates to the rules for shifts in the currency composition of central bank holdings, so as to avoid disruptive movements in exchange rates. Limitations on such shifts, including within existing holdings of reserves, should be adopted, with prior consultation among major currency nations to allow others to take offsetting actions for maintaining greater stability in currency markets.

The third category of disciplines relates to how official reserve holdings are invested. The recent trend toward investment in equity markets through sovereign investment funds and a projected explosion of such investment by China and others opens a whole new area of policy not covered within the existing international economic system. It is addressed here principally in Chapter 9 as an issue of international investment policy. The relation to IMF reform is that sovereign investment threatens to unleash a whole new purpose for reserve accumulation, namely for investment by governments in foreign private companies of strategic interest. The combination of currency manipulation to gain an unfair competitive advantage in trade and strategic foreign investment interests, in fact, could pose the biggest obstacle to IMF reforms related to official reserve holdings.

Linkage between the financial and trading systems.—This linkage already exists through IMF Article IV provisions not to use exchange rates to gain an unfair competitive advantage in trade and GATT Article XV prohibiting the use of exchange rate policy to negate the balanced market access for exports obtained in trade negotiations. The linkage is made explicit through the Article XV provision that the WTO should accept all findings of statistical fact provided by the IMF relating to foreign exchange. The linkage has never been operational, however, although the issue was raised, in broad terms, at the outset of the Uruguay Round negotiations in 1986, when the EU proposed a "balance of benefits" market access condition for chronic trade surplus countries, targeted principally on Japan. The United States blocked that EU proposal, however, in large part because it preferred bilateral negotiations with Japan for sector-by-sector trade balance targets. The currency manipulation issue in recent years, in any event, is far broader in scope in linking exchange rate and trade policies, and now needs to be addressed.

The financial-trade linkage reform would be to clarify and specify the operational content of GATT Article XV, in the same manner as proposed for IMF Article IV, including the IMF statistical linkage already specified in Article XV. The strengthened IMF surveillance procedures for currency misalignment, adopted in 2007, would be the starting point for linking a finding of currency manipulation within the IMF to violation of GATT Article XV commitments within the WTO. No changes would be necessary in the language of Article XV, which would be nigh impossible in view of the WTO unanimity rule for such change. The clarification and interpretation of the role of Article XV could proceed on a plurilateral basis, that is through agreement among the principal trading nations as to how Article XV should be interpreted and pursued within WTO dispute procedures. The critical dimension of the financial-trade linkage is that violation of currency manipulation provisions of the IMF would be subject to WTO dispute procedures.

Six Other Issues

The IMF lending program.—This program would continue to be available to members, but the era of frequent large loans is over, and IMF loans would play a much smaller role in the world of international finance. Moreover, for the many mostly smaller least developed countries, IMF loans of relatively short term at close to market rates of interest are not appropriate. The World Bank is the more appropriate agency for providing longer term, highly concessionary loans related to longer term development objectives. Movement in this direction is already under way.

Macro policy consultation.—This will also continue to be a useful function of the IMF, although as exchange rates become more flexible, the priority for international macro policy coordination diminishes. The 50-year Bretton Woods Commission discussed the desirability of IMF commitments by members for macro policy, but nothing came of it, and the case for doing so is weaker today. An issue for review is how this consultation takes place, including the IMF role in it. There are now three levels of consultation, which can occur on successive days during IMF ministerial meetings: the G-7 extended grouping, the Bank-Fund executive committee, and the Bank-Fund plenary session. The G-7 grouping is the most useful and practical, but it is not a formal part of the IMF structure. Some consolidation, retaining the informality of a smaller G-7 extended grouping, but linked to the IMF, would be useful.

Currency unions and swap arrangements.—Currency unions, such as the European Monetary Union (EMU), and currency swap arrangements, such as the Chiang Mai Initiative among East Asians, are permitted by the IMF and should not be a source of problems. Some clarification of currency union participation may be necessary in terms of authority for dealing with the three major issues described above. Within the Eurozone, for example, the relationship between the EMU board and member finance ministers is not clear for addressing disciplines on central bank intervention, while exchange rate policy related to GATT Article XV would be handled by the EU Commission, which does not currently have a mandate for dealing with exchange rate policy. As for currency swaps and regional reserve pooling in Asia, the most likely issue that could become a problem would be the linkage of these credits, especially if on a concessionary basis, to the exports of the creditor nation. This, however, is a trade rather than a financial policy question, the "tied aid credit" issue, and is left for Chapter 9.

Technical assistance for developing countries.—This useful IMF function would continue, and perhaps be expanded as more of the smaller, poorer countries open their economies to private sector-financed trade and investment. For the growing ranks of the newly industrialized, such technical assistance should be on a fee basis.

International financial market regulation.—This is a technically complex area in need of careful review to see how the IMF could play the most useful role.

Liquidation of gold holdings.—IMF gold holdings, as well as U.S. and other national gold holdings, have far outlived any useful function within the international financial system, while incurring security costs of storage. A basic reform of the IMF provides an opportunity to adopt a schedule for the gradual sell-off of official gold holdings. How the proceeds from such IMF gold sales would be distributed, whether as refunds to contributing members or to pay IMF staff salaries, leads to the issue of organizational reform.

IMF Organizational Reform

Organizational reform should flow freely and logically from the foregoing substantive reforms, but of course it would not. Organizations have great institutional inertia with strong internal vested interests, and this is especially the case for international organizations that are not responsible to profit-maximizing shareholders. The IMF is also politicized in many of its organizational aspects.

In this context, only brief comments are offered here as to what should or might happen. The size of the organization, in any event, should be downsized substantially. IMF staff doubled in the 1990s, related to the surge in IMF loans, and the current staff of 3,000 could perhaps be cut back by at least half.

Voting needs to be revised. It should remain on a weighted basis, somehow related to international trade and investment flows, but certainly not to official reserve holdings. Voting rights, except for the poorest countries, should be linked to payment of budget assessments. Selection of the managing director should be through open election, based on such revised voting shares.

The clear and most important implication of even these very general comments on organizational reform is that the Asian role in the IMF will rise greatly relative to the long-standing dominant roles of Europe and North America. And within Asia, China, and, to a growing degree, India, will have the largest roles, along with Japan and South Korea. How this will work out in practice is unclear. Based on recent currency manipulation in Asia, a much larger Asian role in IMF management can be seen as very threatening, much like the organizational decline of the WTO discussed in the following chapter. But, as the Chinese say, these are interesting times. The only corollary offered here is that the most interesting part of these interesting times in international finance derives from the principal theme of this study—the rise of China and India to become advanced technology superstates, and thus, by definition, "of great financial stature . . . and financially powerful in international affairs."

CHAPTER 9

International Trade and Investment Policy

The world trading system is in disarray, and it is not clear where it is headed. The dominant thrust of trade policies has been toward free trade, and trade continues to grow much faster than GDP in almost all countries. The reduction of trade barriers over the past decade, however, has been principally unilateral and through bilateral free trade agreements (FTAs), while multilateral trade liberalization through the WTO Doha Round has been at an impasse. The advanced industrialized grouping, moreover, which has always played the leadership role in reducing trade barriers, faces rising protectionist pressures in response to growing trade deficits with the newly industrialized grouping.

Within this overall mixed setting, trade relationships among the United States, China, and India have been very different in content and often in conflict. The U.S.-China relationship has been preponderantly bilateral, deeply engaged in negotiations over a wide range of issues, involving a number of serious differences. The U.S.-Indian relationship, in contrast, has focused on the multilateral WTO Doha Round, where they are two of four members of the core negotiating group, although again operating more in conflict than in concert. The China-India relationship is moving cautiously toward more open trade, but with growing Indian concern about the rapid increase in manufactured imports from China. All three are engaged in FTA negotiations, but not with each other.

In these wide-ranging and largely troubled circumstances, an assessment of U.S. trade relationships with China and India, including recommendations for the period ahead, requires a statement of overall U.S. trade strategy, within which the individual bilateral issues can be pursued and related to one another. A frequent format for doing this, and the one adopted here, involves a three-track configuration—the multilateral WTO, the regional FTA, and the bilateral tracks. Unfortunately, however, the U.S. three track trading strategy, as of 2007, is ill-defined, to say the least, especially if based on the degree of bipartisan agreement necessary to implement specific policy objectives.

The presentation here, therefore, begins by summarizing a proposed global three track U.S. trade strategy over the coming two to five years, which is elaborated in the annex. This is followed by track-by-track assessments and policy recommendations for China and India. The remaining sections address three areas of international economic policy that are of growing importance and extend beyond the trade policy agenda, while being deeply related to trade, and where China and India are pivotal participants: international investment policy, development assistance as related to trade, and energy and the environment.

A Proposed Three Track U.S. Global Trade Strategy

A central objective of U.S. trade strategy should be multilateral free trade for nonagricultural trade, which accounts for over 90 percent of total merchandise trade. Parallel objectives would be a more modest degree of liberalization for agricultural trade, a broadening of market access for trade in services, and perhaps a multilateral agreement for international investment. Commitments in the nonagricultural free trade agreement would be plurilateral, that is with limited participation, subject to a threshold of participation by countries accounting for at least 90 percent of total nonagricultural trade. This threshold could be met through participation by the advanced and newly industrialized groupings, while a large number of mostly small, least developed countries would receive free market access for their exports while proceeding at their own pace toward free trade. China and India would have to be participants in such an agreement, and, in fact, they would be decisive: if they opted in, the agreement would succeed; if they opted out, it would fail.

The mutual economic gains from such a free trade agreement would be large for all, especially when full account is taken of the "dynamic" gains from trade generated by new investment and cross border transfers of technology. Such an agreement would also produce mutual benefits from the phase-out of trade-distorting preferential tariffs in FTAs and rules of origin. On balance, U.S. exports should do well because, as in recent bilateral FTAs, other countries usually start with much higher tariffs and therefore make larger absolute reductions in going to zero.

This multilateral free trade agreement could be concluded in 5 or perhaps 10 years, but in any event it would be a two-stage, building-block process. The first stage would be to continue the negotiation of bilateral and group FTAs in various regions, building toward the 90 percent threshold. The trans-Pacific region would be the most important, and the United States should place top priority on further FTAs across the Pacific, in keeping with the regional free trade objective agreed to within the Asia-Pacific Economic Cooperation (APEC) framework. Assuming the U.S.-Korea FTA is approved by the Congress, the next FTAs could be made with Japan, Malaysia, Thailand, Vietnam, and Taiwan. India would be welcomed into the

APEC free trade framework as it proceeds with its FTA with the ASEAN grouping and others, including the United States.

The second stage would be to integrate the various FTA building blocks into a multilateral agreement within the WTO, as described above. Agreement by APEC members and the EU to do this would come close to reaching the 90 percent threshold.

The main reason for this two-stage approach is that neither the United States nor China is prepared at this time to negotiate or seriously discuss free trade with the other. Such a point could be reached in a few years, as the currency misalignment and various bilateral trade issues are worked out. Meanwhile, the momentum of other bilateral FTAs, especially in the Asia-Pacific region, would point toward inevitable U.S.-China inclusion and help create the conditions for achieving this.

The other areas for multilateral negotiation—agricultural trade, trade in services, and investment—would move forward in the most promising and practical setting. Initiatives within the WTO would probably work best on a plurilateral basis. The negotiation of multilateral free trade for the predominant nonagricultural sector should have positive optical and political impact for forward movement in these other areas as well.

There is finally the bilateral track, which will continue to be a major part of trade relationships for dealing with many specific issues, both problem issues and opportunities for more targeted open trade and investment of particular interest to the two parties. The most heavily engaged and problem-oriented U.S. bilateral relationship by far, at least for the next several years, will remain that with China.

This U.S. global trade strategy assumes that the Congress renews the Trade Promotion Authority (TPA), which provides the president authority to negotiate the agreements proposed for the multilateral and regional FTA tracks. If Congress does not renew it, U.S. trade policy will be limited essentially to bilateral issues within existing agreements, while FTAs and broader multilateral agreements will go forward by others, but excluding the United States.

U.S. Trade Strategy Ahead for China and India

Within the proposed three track U.S. trade strategy, China and India will play principal roles, although of a very different character. There will also likely be significant changes in the trade relationships over the coming two to five years. U.S.-Indian trade will become relatively more important, particularly in business services, manufactures, and trade-related investment. The current central focus of the China relationship on the bilateral track and the Indian relationship on the multilateral WTO track should also broaden, perhaps greatly. In any event, all three tracks are in play and need to be interrelated. This is done here, first for China, and then for India, followed by a concluding commentary on the trade policy role of the private sectors.

A Three Track Trade Strategy for China

There has never been a bilateral trade relationship more broadly structured and intensely negotiated than the current U.S.-China relationship. China applied for WTO membership in 1986, which led to 15 years of negotiations and final accession in December 2001. The negotiations involved market access commitments for goods and services, wide-ranging rules to govern trade with China, and objectives related to more than 20 existing WTO agreements. The lengthy negotiating process involved multilateral discussions within a WTO working party and bilateral negotiations with individual working party members, prominently including the United States.

The years since the 2001 Chinese accession to WTO membership have likewise entailed a comprehensive consultative and negotiating process over implementation of the commitments contained in the accession agreement. This process has been far more detailed than the WTO accession negotiations, and is directed at wide-ranging specific issues of technical and legal complexity. It, too, is being carried out at the multilateral and bilateral levels. The WTO has a ten-year transitional review mechanism for monitoring compliance, with annual status reports. Bilaterally, the U.S.-China Relations Act of 2000 requires the U.S. Trade Representative (USTR) to report annually to Congress on Chinese compliance, based on an oversight compliance program, carried out through the ministerial level bilateral Joint Commission on Commerce and Trade (JCCT). The agenda and activities of this bilateral mechanism are comprehensive and include consultations with thousands of U.S. companies.

The 108 page 2007 USTR Report to Congress examined nine broad categories of WTO commitments by China, within which there were 32 more specific topics. The principal conclusions of the report were:

> "China has taken many impressive steps to reform its economy, making progress in implementing a set of sweeping commitments that required it to reduce tariff rates, eliminate nontariff barriers, provide national treatment and improved market access to goods and services imported from the United States. . . . China's implementation of its

> WTO commitments has led to significant increases in U.S.-China trade. . . .
>
> "Through the first few years after China's accession to the WTO, China made noteworthy progress in adopting economic reforms that facilitated its transition toward a market economy. However, beginning in 2006 and continuing throughout 2007, progress toward further market liberalization began to slow. . . . China's difficulties in generating a commitment to the rule of law have exacerbated this situation.
>
> "At the root of many of these problems is China's continued pursuit of problematic industrial policies that rely on excessive Chinese government intervention in the market through an array of trade-distorting measures. . . .
>
> "Evidence of a possible trend toward a more restrictive trade regime appears most visibly in a series of diverse Chinese measures over the past two years signaling new restrictions on market access and foreign investment in China. . . ."[136]

Clearly much more needs to be done for China to reach full compliance with its WTO obligations, and U.S. policy should be to press with increased vigor to reverse the regressive tendencies noted during 2006 and 2007, and to attain full Chinese compliance. A detailed assessment of next steps is not attempted here. Rather, broad comments are offered on four aspects of the general direction and orientation of the bilateral negotiations.

The heavy concentration on issues relating to advanced technology industry.—Such issues recur throughout the USTR report. The number one priority is protection of intellectual property, which involves billions of dollars of annual losses to U.S. firms and goes to the heart of U.S. leadership in technological innovation. The report is especially critical of Chinese performance on IPR: "China has continued to demonstrate little success in actually enforcing its laws and regulations in the face of the challenges created by widespread counterfeiting, piracy and other forms of infringement. . . . Legal measures in China that establish high thresholds for criminal investigation, prosecution and conviction preclude remedies in many instances of commercial-scale counterfeiting and piracy, creating a 'safe harbor' for infringers" (page 76). "Trade-related investment measures are another cause for strong concern. China agreed . . . 'to eliminate WTO-inconsistent

---

[136] United States Trade Representative *2007 Report to Congress on China's WTO Compliance*, December 2007, pp. 3-4.

requirements relating to export performance, local content, foreign exchange balancing and technology transfer.' However, six years after China's WTO accession, some of the revised laws and regulations continue to 'encourage' technology transfer. . . . U.S. companies remain concerned that this 'encouragement' in practice can amount to a 'requirement'" (page 60). Other areas of the report related to advanced technology industry include subsidies, biotechnology regulations, telecommunications services, and a lengthy ten-page section on standards, technical regulations, and conformity assessment procedures.

The new Chinese industrial policy of indigenous innovation.—This recent policy direction, discussed in Chapter 2, is working its way into the bilateral trade negotiations and is a principal reason for the USTR conclusion that progress toward market liberalization slowed since 2006. The report cites an American trade association which observes: "Designation of 'pillar' industries, promoting 'indigenous innovation,' and establishing 'national economic security' criteria to review deals are troublesome signposts that do not imply full market access for U.S. companies" (page 6). The USTR Report raises concerns, in this context, about the July 2005 *Steel and Iron Industry Development Policy*, the June 2006 *State Council Opinions on the Revitalization of the Industrial Machinery Manufacturing Industries*, the August 2007 *Anti-Monopoly Law*, and the November 2007 revised *Sectoral Guidelines Catalogue for Foreign Investment*" (pages 61-63).

Greater use of the WTO dispute mechanism.—China has been more responsive to complaints pursued through WTO dispute procedures, where there is a strong case made against Chinese practices. The U.S. was slow to use the WTO dispute mechanism while China was still at the early stages of implementation of its WTO obligations. Now, with the implementation schedule almost complete, the WTO option should be and is being used more frequently. The first U.S. complaint filed against China in the WTO was in 2004, over value added taxes that discriminated against imported semiconductors, and China complied. Two cases in 2006 involved Chinese anti-dumping duties on imported kraft liner board and discriminatory charges on imported auto parts. China dropped the anti-dumping duties before a formal complaint was made, and the auto parts case remained in progress during 2007. The U.S. brought three more cases to the WTO in 2007, involving subsidies, IPR, and restrictions on copyright-intensive industries. It is noteworthy that the auto parts complaint was filed jointly by the United States, the EU, and Canada, which can be more effective in building a case for China to take remedial action. Such joint complaints can and should become more frequent as the Chinese trade surplus with Europe and other regions grows and creates political pressures for governments to be more active in enforcing China's WTO obligations.

The indirect political impact of the surging Chinese trade surplus.—The political pressures generated by the surging Chinese trade surplus may not only lead to greater use of WTO dispute procedures, but may also affect the broader course of U.S.-Chinese bilateral trade negotiations within the JCCT. These political level meetings are publicized, in part, as the means for positive steps in opening the Chinese market to U.S. exports, reducing the U.S. trade deficit with China, and thus countering the protectionist pressures against China in the United States. In this context, the U.S.-China trade deficit can provide a strong argument for China to be more responsive. The shortcoming of this route, however, is that even significant positive actions by China will have relatively small impact on a U.S. bilateral deficit of over $200 billion a year and rising. A decline in the trade deficit depends far more on exchange rate adjustments than on individual trade policy actions taken within a highly complex and slow-moving Chinese trade bureaucracy.

These are the broad lines of the dominant bilateral dimension of the U.S.-Chinese trade policy relationship. The multilateral and regional FTA tracks are of far less consequence, at least for direct dealings between the two countries, but this should change somewhat over the next several years, especially if the free trade strategy as proposed here is pursued.

The multilateral WTO track has been bogged down for six years in the Doha Round negotiations to increase market access for exports. China, while rising to become the number one exporting nation, has stayed on the sidelines and has played no significant leadership role. The Chinese role will likely increase at the closing stage of the Round, however, whether the final result is a modest "success" or an outright failure which could be couched in terms of an indefinite suspension. A modest success would require significant reductions in trade barriers by China, particularly in view of its large and growing trade surplus, and could involve difficult final negotiations with the United States, the EU, and others. Full reciprocity from China should be sought, while the Chinese position has been that it does not need to provide reciprocity because it is a developing country. This position should be challenged, especially by the United States, which did not accept Chinese status as a developing country at the time of WTO accession.

A more important dimension of the multilateral track could come to the fore at the close of the Doha Round, no matter how it ends, when members seek to agree on next steps for the organization. This has been standard practice in past rounds. The previous Uruguay Round final agreement included a work program for international investment, competition policy (including anti-dumping duties), and trade facilitation, such as customs clearance, with a view to possible future negotiations, although only trade facilitation made it onto the Doha Round agenda.

This time bigger questions will be raised as to what happens next in each major sector of trade, after what, in any event, will be judged a disappointing Doha Round experience. Certainly the role of China, as well as India, will be critical for establishing a post-Doha Round WTO agenda. The agenda will likely have to respond to the proliferation of FTAs and to the nonproductive, heavily mercantilist orientation of the Doha Round negotiations, as explained below in the annex. The big option of multilateral free trade for nonagricultural trade, as proposed here and as was officially proposed by the United States in the opening stage of the Doha Round, should at least be discussed, and it could be included in the post-Doha Round work program for analysis and possible future negotiation. More precisely, if the United States, China, and India were to support such a WTO work program proposal, it would be adopted, and the final assessment of the Doha Round would be significantly more positive with respect to the future of the WTO.

This outcome on the multilateral track would stand the best chance of success if pursued in parallel with actions on the regional FTA track and, in particular, within the APEC free trade framework, as the building block first stage toward the multilateral free trade objective. The U.S.-China relationship will again be decisive in how the Asia-Pacific regional free trade course unfolds, even though direct free trade negotiations between the two countries would not be feasible for at least a few years. A three-pronged U.S. approach is proposed here as a critical part of the overall global free trade strategy:

First, the United States should press for reaffirmation of the APEC free trade objective, with a 2020 date for implementation. China would likely be the most reluctant member, and the United States should make this issue a priority for discussion in the Strategic Economic Dialogue and in summit meetings.

Second, the United States should pursue further FTAs across the Pacific, with Japan, Malaysia, Thailand, Vietnam, and Taiwan. Such agreements, together with existing agreements with Australia and Singapore, and with South Korea assuming congressional approval, would make the Asia-Pacific scope of regional free trade, including the United States, close to a *fait accompli*, which China, however reluctantly, would have to accept.

Third, the United States should propose, again within the Strategic Economic Dialogue, that the United States and China establish a bilateral experts group to do a cost-benefit analysis of U.S.-China free trade within the broader APEC free trade framework. One precedent for such a step is the China-South Korea experts group. The economic results of such an analysis are almost certain to be positive in terms of the mutual gains from trade, and the existence of an analytically sound report should set the stage for preliminary discussions between the United States and China as to how the two countries could pursue both the APEC and, ultimately, the WTO free trade objectives.

A Three Track Trade Strategy for India

Since 2001, there has been a stark contrast between the U.S. three track trade strategies with China and India. In December 2001, as recounted above, China acceded to the WTO, triggering six years of intensive bilateral track negotiations. A month earlier, in November 2001, the WTO Doha Round was launched, within which the United States and India, together with the EU and Brazil, came to play the inner-circle leadership role for this top-priority multilateral track initiative. The result for the U.S.-India trade relationship has been that media and trade policy establishment attention in Washington have focused almost exclusively on the continuing Doha Round impasse, highlighting the conflict between minimalist Indian offers to reduce its trade barriers and more ambitious U.S. objectives. The response of trade policy experts when asked about bilateral trade issues with India is usually a blank stare.

This U.S.-India trade policy orientation, however, is likely to change significantly over the coming two to five years in view of the rapid growth in U.S.-Indian trade, and in the context of a likely reformulation of Indian trade strategy which, during 2001-2007, can be characterized as having been in consummate contradiction. This harsh assessment requires some explanation.

For decades India was one of the most protectionist members of the GATT, now incorporated into the WTO. As recently as 2000, India still had the second highest average tariff level, at 31 percent, after Pakistan, out of more than 100 WTO members, compared with 7 to 17 percent average levels for East Asian nations, including China. India also took a highly protectionist, minimalist position during GATT Rounds of negotiation to reduce import barriers. The 1986 GATT ministerial meeting in Punta del Este that launched the Uruguay Round, for example, pitted the group of 48 (G-48) industrialized and developing countries, which had expanded to 60 by the time of the meeting, against the "hard-line" G-10, led by India and Brazil and including Cuba, Nicaragua, and Egypt. Within Asia, India was the only member of the G-10, while 10 other Asians were part of the G-48. The G-48 proposed an ambitious agenda for large reciprocal reductions in trade barriers and the extension of the trading system to include trade in services, trade-related investment measures, and protection of IPR, whereas the G-10 proposed much smaller reductions and opposed all three extensions. The G-48 strategy was to hold firm and negotiate intensively with the more reasonable Brazilians, while isolating India. As India became more isolated, the G-48 prevailed almost entirely, and the Uruguay Round was successful in achieving all its major objectives, despite protracted Indian opposition during the course of the negotiations on many issues.[137]

The radical change in political orientation within the Doha Round is that India and Brazil, the leaders of the hard-line minimalist G-10 in the previous round, became the designated negotiators for the entire developing country grouping of over 100 participants. This goes far to explain the six-year impasse and disappointing prospects for the final outcome, as explained in the annex. And again a distinction has existed between Brazilian and Indian performance. Brazil has held out for greater liberalization of agricultural trade before being more forthcoming on nonagricultural and services trade, while India has offered minimal reductions in its trade barriers simply on the grounds of being a developing country.

This long-standing, highly protectionist Indian performance within the GATT and the WTO, in Geneva, contrasts fundamentally with recent Indian trade policy at home, involving major unilateral reduction in trade barriers to imports and investment, including the establishment of essentially free trade SEZs, and the negotiation of FTAs as important parts of its market-oriented, open-trade economic reform strategy. Nonagricultural tariffs, on average, remain high, but many are now down to 10 percent, and some were reduced further in 2007 to contain inflationary pressures. Trade in services is being progressively liberalized and is very open for telecommunications and business services. Protection of intellectual property is being strengthened, partly in response to the interests of technology-oriented Indian firms.

This is the essence of the consummate contradiction label given to recent Indian trade policy. How India's trade policy is likely to move toward greater consistency in the market-oriented, open trade direction and how the U.S.-India trade relationship can play a significant role in this transition can best be presented track by track.

The U.S.-India bilateral relationship since 2005 has focused heavily on the civilian nuclear energy agreement, which has significant economic content related to the energy sector, but predominantly concerns national security and foreign policy interests—the impact of the agreement on nuclear nonproliferation, Indian sovereignty, and a strengthened U.S.-

[137] See Ernest H. Preeg, *Traders in a Brave New World: The Uruguay Round and the Future of the International Trading System* (University of Chicago Press, 1995). The Punta del Este meeting is recounted on pages 1-10.

India political/military relationship in the face of rising Chinese military power. As for the trade relationship per se, as noted above it has centered on the Doha Round impasse, with relatively little attention given to bilateral track issues. Some positive bilateral steps to facilitate trade and investment have been taken, however, and a more active bilateral agenda should be pursued, while dealing separately with basic differences within the WTO.

A landmark bilateral step was the 2005 Open Skies aviation agreement, which permits unlimited flights for the airlines of both countries, including service by cargo carriers for in-country flights without direct connection to the homeland. This agreement is truly open skies and far surpasses the highly restrictive U.S.-China air agreement.

Active bilateral discussions are being pursued for opening the Indian market to trade in services, particularly financial and retail marketing services, which are moving forward, however slowly, in tandem with the domestic economic reform agenda. One result, in keeping with the Indian strategy of more liberal interpretation of existing regulations, was the July 2007 50-50 percent joint venture between Wal-Mart and the Indian retail chain Bharti Enterprises. Direct sales by foreign multibrand retailers are still not permitted, but this joint venture, with Wal-Mart in a supportive role, particularly for its logistical capabilities, was found acceptable and is a significant step toward more open international competition, with retail marketing being progressively liberalized for Indian firms.

An example of how the WTO dispute procedure can be used to open the Indian market on the bilateral track was the May 2007 U.S. request for a dispute panel to seek elimination of Indian "additional duties," ranging from 20 percent to 150 percent, on imports of beer, wine, and spirits. In July, faced with a weak case before the panel, India announced the withdrawal of the additional duties which, if duly implemented, would benefit not only California wine and Kentucky bourbon exporters, but also burgeoning Indian middle-class consumers who now face extraordinarily high-priced wine and spirits.

In the other direction, the number one Indian concern in the U.S. market is the continued ability of skilled Indian workers to obtain H-1B visas for work in the United States with Indian companies. Ten of the 20 top users of H-1B visas in 2006 were Indian companies, concentrated in the software and business services sector. Congressional concerns about the outsourcing of U.S jobs could lead to a reduction in the current 65,000 annual quota for such visas, while Indian and high-tech American companies want to see an increase to over 100,000. Indian official statements about the negative consequences of a cutback, along with strong support for increases by the Indian-American community, including generous campaign contributions to prominent members of the Congress, should help maintain the liberal policy for H-1B visas which, in any event, has broad U.S. public support.

These are examples of the bilateral trade track through 2007, linked largely to the economic reform program being implemented within India. There is broad agreement to forge closer trade and investment ties as a mutual interest. In November 2006, Under Secretary of Commerce Frank Lavin and 250 U.S. corporate leaders met with their counterparts in an "India Business Summit" in Mumbai. The challenge is to convert this high-level political commitment to a more specific, action-oriented bilateral track agenda.

The regional FTA strategies of the United States and India, including a possible bilateral FTA, are in a period of gestation in both countries, which is a good time to begin serious discussion of longer term mutual interests. The United States needs renewal of Trade Promotion Act authority to negotiate additional FTAs, which would be targeted most importantly on trans-Pacific FTAs. India has a high-visibility and popular FTA with Singapore, and is engaged in slow-moving negotiation for FTAs with ASEAN and the EU. Both countries are thus poised to be major participants in the proposed stage one, building block strategy toward multilateral free trade. This strategy, moreover, could and should become integrated through two far-reaching decisions.

The first decision would be for India to join APEC, with the strong support of the United States, including India's explicit acceptance of the regional free trade objective. India is a major Asian trading nation, and it is increasingly anomalous for regional free trade to be pursued without Indian participation.

The second decision would be for the United States and India to begin preliminary discussion for a bilateral FTA. Some informal talks have already taken place, both officially and between private sector organizations. As with other recent U.S. FTAs, the objective would be total free trade for the nonagricultural sector, with somewhat longer phase-in periods for a few problem industries, more modest liberalization for agriculture, and broad provisions for trade in services, investment, IPR, and other issues. Such an agreement is highly feasible in technical terms, in view of its compatibility with the current path of Indian economic reform, would bring large mutual economic benefits, and would provide a stable, high-visibility foundation for a strengthened overall relationship between the two largest democracies, committed to economic freedom and free trade. A bilateral experts group to undertake a cost-benefit analysis of bilateral free trade, as proposed for the U.S.-China relationship, should have even greater appeal for a forward-looking U.S.-India relationship.

Such initiatives on the regional FTA track would have significant consequence for the future course of the multilateral WTO track as well. Preliminary discussion of a U.S.-India FTA, presumably going forward in parallel with continued EU-India FTA negotiations, could be decisive in generating serious discussion within the WTO of the objective of consolidating the now hundreds of FTAs within a multilateral free trade agreement. Sponsorship by both the United States and India for WTO study of the objective, with a view to possible future negotiation, would almost certainly ensure its adoption. Such a multilateral free trade study program would be a far cry from long-standing Indian positions within the WTO, but the Indian open-trade economic reform strategy in 2008 is an even further cry from the protectionist policy India took 20 years ago, at the time of the Punta del Esta meeting.

Private Sector Trade Policy Roles

The proposed three track U.S. trade strategies for China and India over the coming two to five years have been presented in terms of interaction and negotiation among the governments. There has also long been, however, an important role for private sectors in the formulation of trade policy which, on balance, has supported more open trade and investment. This private sector role has been predominantly within the advanced industrialized grouping—the United States, the EU, Japan, and Canada, in particular. But now, given everything presented in this study, the Chinese and Indian private sectors are of growing influence within their respective economic power structures, and the question is how this influence will be exercised in shaping trade policy.

Private sectors in the industrialized grouping have often been out in front of governments in terms of trade policy initiatives. In the Uruguay Round, the U.S., European, and Japanese private sector organizations together successfully pressed their often reluctant governments to put trade in services and intellectual property rights on the negotiating agenda. For tariff reductions, they convinced their governments to give priority to free trade or low common tariff levels by sector, as more fruitful than complicated tariff-cutting formulas on which, as in the Doha Round, governments were unable to agree. In the mid-1990s, the private sector organizations took the lead to convince their governments to negotiate an international investment agreement within the OECD, which came close to fruition before environmentalist groups forced the United States to withdraw. The role for private sector leaders in the Doha Round has been smaller because of the overriding priority given to agriculture and the impasse over formulas to reduce tariffs and barriers to trade in services.

The rapid growth of trade-oriented Chinese and Indian companies points in the direction of common interests with industrialized country private sector organizations. Protection always gives an advantage to domestic producers, but as tariffs come down to relatively low levels and corporate strategies involve ever deeper cross-border supply chain management, the advantages of predictable open markets become larger. Free trade agreements have special advantages, not only in eliminating import duty payments and most customs processing costs, but also in providing the best guarantee that export markets will remain open, which can be decisive when firms are making long-term investment decisions. This was clearly the case for U.S. and Mexican firms within NAFTA, wherein largely investment-driven trade in both directions tripled in a decade.

Chinese and Indian private sector organizations are beginning to have contacts with American, European, and Japanese counterparts, but with India well in the lead. Chinese business organizations have very limited freedom to collaborate with foreign business organizations, especially if collaboration involves criticism of official Chinese trade policy. Indian organizations, in contrast, are fully independent and highly active in pressing their government to change policies, usually in the direction of more market-oriented and open-trade economic reforms. The Confederation of Indian Industry (CII), the Federation of Indian Chambers of Commerce and Industry (FICCI), and the National Association of Software and Service Companies (NASSCOM) are all heavily engaged in the policy process, including through Washington offices by the CII and FICCI.

For decades, the organizational center for much of the private sector dialogue and coordination on trade and investment strategies has been in the OECD Business and Industry Advisory Committee (BIAC). BIAC has a full agenda of trade, investment, and technology policies, as well as of specific issues which, during 2007, included structural economic reforms, international taxation, consumer protection, and corruption. Within the OECD, the Indian private sector is not only ahead of the Chinese private sector, but of the Indian government as well. CII joined the BIAC in 2006, where nonmembers are permitted to participate as "observers," and has been particularly active in BIAC discussion of international investment. The Chinese private sector, however, has still not been able to put forward representation which is fully independent of the government, which is a BIAC requirement. The Indian government, meanwhile, still resists official participation in the OECD, on the foreign policy grounds that this is the rich man's club and might imply Indian reciprocity in trade negotiations, even while China and a number of other nonmembers have long officially participated in selected OECD activities, especially the work of the Directorate of Science, Technology, and Industry.

An important issue related to the presentation here is whether Indian and Chinese private sector organizations will play an influential role in the development of FTAs with the United States, and ultimately free trade on a multilateral basis within the WTO. It should be in the longer term interests of both Indian and Chinese multinational companies to move in this direction, which would help ensure permanent free access to the U.S. market. Again, the Indian private sector is well ahead of its Chinese counterpart. Informal positive talks have taken place between CII and the U.S. National Association of Manufacturers (NAM) about a bilateral FTA, at least for manufactures. NASSCOM should also be a strong advocate of more open trade for software and business services, bilaterally and multilaterally.

Thus, the outlook is for divergent courses ahead for Chinese and Indian private sector participation in the formulation of trade policy. Chinese organizations will have far more limited freedom of action, in addition to having to confront the difficult restructuring to less export- and more domestically oriented growth, as discussed in Chapter 5. Indian private sector organizations, in contrast, are both more independent and more self-confident as global market competitors. This should lead to deeper Indian dialogue and collaboration with American, European, and Japanese counterpart organizations, as well as a growing role in influencing Indian trade policy. Indeed, a forward-looking, free trade-oriented Indian private sector may come to play an important role in changing the backward-looking, protectionist-oriented official Indian representation in the WTO.

International Investment Policy

International investment and trade are becoming more deeply integrated in policy as well as in market terms, but investment policy is only included in small part in the WTO multilateral trading system. Rather, hundreds of bilateral investment agreements and treaties have accumulated over the decades, and more recently investment has been included in many FTAs, most comprehensively in U.S. FTAs. These agreements usually cover the right of establishment, equal or national treatment, compensation for expropriation, and dispute procedures. The possible consolidation of these many bilateral agreements into a multilateral framework is addressed in the annex to this chapter.

The future course of investment policy, however, faces two important new challenges stemming from major developments under way in the pattern of international investment. Both challenges involve investment by public sector, as distinct from private sector, investors, that is, investment by government, or "sovereign" enterprises and funds. China is emerging as the central actor in both cases, in quantitative as well as political terms. Both challenges, moreover, are in largely uncharted policy waters.

The first challenge relates to direct investment, including acquisitions, undertaken by state-owned or controlled enterprises. Under existing bilateral investment agreements, distinctions are not generally made between private and public sector investors, but there can be important differences in motivation. Private sector investment is presumably driven by market forces and profit potential, whereas state-owned enterprises might have other objectives. For example, Chinese state-owned companies may seek to acquire foreign energy and advanced technology companies as part of a broader economic strategy, and will thus be willing to pay a higher, in effect, subsidized price than would profit-motivated private investors. The WTO trading system has distinct criteria for state-owned enterprises in the application of its market-based rules for trade, and it prohibits export subsidies, under the threat of countervailing duties. Bilateral investment agreements, in contrast, generally do not distinguish between private and public investors, and are silent on the issue of subsidies.

There is always some degree of screening for approval of new foreign investments and acquisitions, as under the U.S. Committee on Foreign Investment (CFIUS) law, whose authority is limited to restrictions on foreign investment related to national security. The CFIUS review process was tightened somewhat by legislation in July 2007, but only about 10 percent of investments are subject to review, and very few are opposed. With the prospect of much larger public sector investments, particularly by China and some oil exporters, the issue raised here is whether a broader screening process should be adopted. Should some R&D-intensive, energy, and infrastructure projects be restricted from takeover by foreign government-owned or controlled companies? The subsidy factor should also be considered, although assessing investment subsidies is more complicated than assessing export subsidies. For example, in 2005 China's Cnooc made what was widely considered an excessively high offer to buy the American petroleum company Unocal, likely because of Unocal's holdings in Southeast Asia, of geo-political interest to China. When Cnooc backed away in response to congressional criticism, Chevron made the purchase, and the issue raised here is whether it would have been unfair for Chevron to have lost the acquisition because of a subsidized bid by state-owned Cnooc. In conceptual terms, the parallel restriction for trade policy is the ban on export subsidies and the threat of countervailing duties to offset prohibited subsidies.

No specific proposal is offered here except that a clear statement of policy should be developed in what is now largely a policy void, and the direction of the policy should be to limit subsidized, nonmarket behavior of foreign state-owned enterprises, particularly for advanced technology and other industries where broader national interests of the investor government are likely to be pursued.

The second challenge for the future course of investment policy is potentially far greater and is into totally uncharted policy waters. This is the issue of state or so-called sovereign investment funds for foreign portfolio investment. The issue derives directly from the rapid buildup of official foreign exchange holdings as a result of currency manipulation and high oil prices, discussed in Chapter 8 as a major problem facing the international financial system. This foreign exchange buildup can also become an investment policy problem when governments with excessive holdings decide to invest large portions of the buildup in equity markets, at a minimum to obtain higher returns than through the traditional practice of holding interest-bearing government securities but also possibly to pursue other policy objectives.

Some countries with large reserve holdings, such as Kuwait, Norway, and Singapore, have had sovereign investment funds for years, but the practice received little attention because the amounts of investment were not extremely large and the management of the funds was carried out on an apparently market-oriented basis. The Norwegian Norges Bank Investment Management, however, is subject to policy screening that excludes stock purchases of weapons producers and companies judged to be serious or systematic violators of human rights and perpetrators of environmental degradation. Wal-Mart, for example, was sold off as a serious violator of human rights, based on criticism by anti-Wal-Mart nongovernmental organizations. The question now is whether other countries, such as China and Russia, will give more important political direction to their sovereign investment funds. This whole issue, moreover, has recently grown enormously in potential size, with official foreign exchange holdings reaching $6 trillion in 2007 and headed for at least $12 trillion by 2015, and with China alone accounting for close to one-third of the total. Moreover, an additional $2 trillion has already been transferred to sovereign investment funds and is no longer listed as official reserves.[138] The two questions addressed here are how much of the $6 trillion rising to $12 trillion will

[138] See Edwin M. Truman, *Sovereign Wealth Funds: The Need for Greater Transparency and Accountability*, Policy Brief (Peterson Institute for International Economics, August 2007). The presentation explains the difficulties in gathering accurate total figures or any detail as to what is happening.

be shifted into the $2 trillion category of sovereign investment funds, and how will governments, and the Chinese government in particular, manage these funds?

China officially entered the sovereign investment fund arena in 2007 with the creation of the China Investment Corp., headed by the former vice minister of finance, Lou Jiwei, with an initial funding of $200 billion, which could possibly rise to $1 trillion or even $2 trillion over the next several years, with total reserves headed to $3 trillion by 2010. This new enterprise, as part of the Chinese government, raises a number of questions about Chinese investment strategy and the policy response by the United States and other recipients of such government-directed investment. How will China allocate the investment by country and region and how will decisions be made for significantly shifting from one region to another? Which will be the priority sectors for investment? Will energy and advanced technology companies receive top priority?

There are also important issues of management influence, including board membership, in companies where foreign sovereign holding reaches the threshold level for such management influence and board participation. The Norwegian fund is explicit that a priority area of its strategy is the exercise of its right to vote, including for board members. This may not be a significant problem for Norwegian-supported board members on a few American companies, but if dozens or more American companies, especially in advanced technology industries, were to be owned in substantial part by the Chinese investment fund, with Chinese official representation on the board, U.S. technological leadership and export competitiveness could be compromised.

These are the new and potentially very large investment policy challenges facing the United States that go beyond the existing investment policy framework. The Chinese sovereign investment fund will probably take some time to get organized and become fully implemented, but it is not too soon for the United States to develop a policy response, which may influence the course of Chinese implementation. The enormity of the potential scope of the issue, however, should be clearly recognized. A projected $1 trillion to $2 trillion of Chinese sovereign investment funding would equate to $1 billion to $2 billion equity holdings in 1,000 companies, of which several hundred are likely to be American. For many of these companies, a $1 billion to $2 billion shareholder interest would entail considerable management influence, including through membership on the board and market leverage from a threatened sell-off.

The principal recommendation made here is that the whole subject receive an in-depth, forward-looking analysis. The following lines of possible policy response are offered not as definitive proposals but as starting points to stimulate such analysis.

A concerted, not unilateral, response is necessary.—Similar if not common approaches would be far preferable as a way to have a balanced impact on markets. The United States should find common ground with Europe (except Norway) and Canada. Germany and France have expressed growing concern about sovereign investment in the energy and other strategic sectors, particularly by Russia and China. Japan may move toward portfolio investment for some of its foreign exchange holdings, but should also be willing to consider reasonable guidelines and regulation, especially in view of likely Chinese sovereign investment in Japan. In any event, a common U.S.-EU approach would be the vital core because, to paraphrase Willie Sutton, that's where the equity markets are.

Separate responses are needed for direct investment, including acquisitions, and sovereign portfolio investment.—The need for separate responses has already been explained, in that direct investment would be subject to the existing screening process, adjusted for the characteristics of greatly expanded state-owned investments, whereas a new framework will be required for dealing with sovereign fund portfolio investment. In this context, an important difference in impact on financial markets is that direct investments are long-term decisions, while portfolio investment is managed on a day-to-day basis, with potentially more volatile short-term impact on markets. The remaining issues for study presented here are thus directed to the more complex situation created by large-scale sovereign fund portfolio investment.

Transparency requirements for sovereign investment funds.—This would be the starting point and should certainly include levels of purchases and sales by national market, on an ongoing basis, and greater detail on a periodic basis. The Norwegian fund does provide considerable transparency, whereas most other existing funds do not.[139]

Limits on national or regional shifts for sovereign investment funds.—This could become a complicated regulatory matter. If China, for example, were to have several hundred billions of dollars of portfolio investment in both the U.S. and European markets, it could, for commercial or political reasons, shift or threaten to shift large amounts

[139] The existing state of transparency is discussed in Edwin M. Truman, "The Management of China's International Reserves: China and an SWF Scoreboard," presented at a conference on China's exchange rate policy at the Peterson Institute for International Economics on October 19, 2007. Truman provides a scorecard for 33 existing sovereign wealth funds for standards and structure, governance, transparency, accountability, and behavior. The Chinese Central Huijin Investment Company, established in 2003 with $67.5 billion to recapitalize four state-owned banks, scores below average.

from one market to the other, causing a disruptive decline in the market being sold off. In view of comments in China, cited earlier, about threatening to sell official dollar holdings related to other policy objectives, and the follow-up statement in November 2007 that China will favor stronger over weaker currencies, a reasonable safeguard could be limits on short-term sales or purchases, at least for the few largest sovereign investors, subject to consultation. The starting point, in any event, would be adequate transparency to observe the market behavior of the funds.

Restraint on voting rights and board membership.—One suggestion in this area has been to set a percentage limitation on sovereign foreign ownership in individual companies, perhaps 10 percent or 20 percent, but in a large, widely held company, these percentages could still garner considerable voting influence, including representation on the board. A broader approach would be to prohibit voting rights and board membership for sovereign investors on the grounds that, with trillions of dollars in play, foreign governments would be able to acquire unwarranted influence in a large number of American—and European—companies. It would not be unreasonable to do this. In effect, it would simply place foreign sovereign investors on the same footing as smaller private investors with insignificant voting rights and influence on management. It is noteworthy, in this regard, that when China made its first major sovereign investment of $3 billion in The Blackstone Group in 2007, Wang Jianxi, the Chairman of China Jianyin Investment, who signed the agreement on behalf of the new investment fund, said that the fund decided to give up its voting rights on its stake in Blackstone because "we deemed it to be a purely financial investment."[140] The prohibition described here would simply require that all purchases by the Chinese and other sovereign investment funds be similarly made as purely financial investments and not for other purposes. Again, a first step could be transparency, for the funds to report their voting behavior and membership on boards of invested companies.

A policy linkage to reform of the international financial system.—The various possible regulatory actions described above for sovereign portfolio investment could be justified not as new regulatory initiatives but as an interim response to the out-of-control central bank purchases of foreign exchange that are related to the currency manipulation issue, purchases that need to be brought under control within the international financial system. The presumption would be that such regulatory actions could be relaxed as excessive official foreign exchange holdings were reduced.

---

[140] *Financial Times*, May 27, 2007.

There is finally the situation of India. It is not clear how India will respond to the extent its $300 billion of foreign exchange holdings continues to increase. In broadest terms, the issue is not whether India invests in U.S. Treasurys or foreign stock markets but whether these hundreds of billions of dollars of national savings should be invested abroad or at home. Certainly the latter alternative is in the interest of the large majority of Indians who live in or close to poverty. If the former foreign investment alternative is pursued, the Indian sovereign investment agency would have to accept the same disciplines as applied to China and other sovereign investors. In any event, the prospect of very large, sovereign foreign investment by China and India is an appropriate transition to the subject of development assistance, and specifically whether official development assistance, in this new age where "developing country" central banks hold trillions of dollars of highly liquid foreign assets, is being overtaken by events.

Development Assistance as Related to Trade

The role of official development assistance (ODA), or economic aid, has been changing greatly as many of the larger developing countries have become creditworthy and no longer need long-term ODA loans, while ODA disbursement has become more concentrated in the poorest countries of Africa and parts of South Asia and the Caribbean Basin. The provision of large amounts of ODA to this grouping of poorest countries raises a number of questions for the multilateral development banks (MDBs)—the World Bank and the four regional banks—and bilateral aid programs, questions involving the need for more concessionary loans and grants, close coordination of aid projects within country development strategies, and standards for good governance. These issues are not addressed in detail here, and ODA is discussed principally in the context of India and China phasing down their roles as ODA recipients and China, in particular, shifting to become a large, and perhaps the largest bilateral aid donor, with important impact on trade.

India remains the largest borrower from MDBs—from the World Bank and the Asian Development Bank—with new loans totaling $1.4 billion and $1.7 billion, respectively, in 2006, or more than 10 percent of total loans by the two banks. These $3 billion of loans, however, pale by comparison with an inflow of FDI projected to approach $30 billion in 2007, other private capital inflow, and exports of goods and services rising above $200 billion per year. Particular aid projects targeted on rural and urban poverty still play a valuable role, including through project technical assistance managed by the banks. The relative role of ODA for India, however, is greatly diminished from what it was 20 or even 10 years ago. The issue of relative need also should be addressed. A basic criterion for MDB loan eligibility, related to the standard central bank guarantee, is whether the borrower is creditworthy for long-term development loans on commercial markets. India, with $300 billion of foreign exchange in the central bank, is now eminently creditworthy, and perhaps at least some of the $3 billion of MDB loans could be more equitably reallocated, on softer terms, to poorer and less creditworthy aid recipients. The same kinds of judgment are being made about bilateral aid to India. The United States reduced its bilateral aid to India by 35 percent, to $81 million, in its 2008 budget request on the grounds that India is now a "transforming" rather than a "developing" country and is therefore less needy than other aid recipients in the region, such as Pakistan and Bangladesh.[141]

As India phases down as an ODA recipient, it will also likely begin to rise as an ODA donor, especially to its poorer neighbors in South Asia, and is therefore encouraged to follow closely the problem-laden rapid rise of China as an aid donor, including interactions with the MDBs and other bilateral donors.

China, too, remains one of the largest aid recipients from the World Bank and the Asian Development Bank, with new loans of $1.5 billion and $0.7 billion, respectively, in 2006. Suffice it to say that this is absurd, bordering on the obscene. With $1.5 trillion in its central bank at the end of 2007, China has no need of $2 billion of long-term loans from the MDBs, which only add to the $1.5 trillion and could be lent to far better purpose elsewhere. China should shift from the borrower to the lender side of the table in both institutions, and receive positive recognition for doing so. If the World Bank can supply China with useful technical assistance, related to poverty elimination and to environmental improvement, such assistance can continue to be provided on a cost basis.

The more important ODA-related issue for China is its rapid rise as a bilateral aid donor. Foreign assistance provided by the China Development Bank (CDB) rose to $8 billion in 2006, and the bank's governor, Chen Yun, stated the intent to increase the program, particularly in the area of natural resource projects by Chinese firms.[142] The Chinese Export-Import Bank, working in close collaboration with

---

[141] *Financial Express*, July 25, 2007. State Department spokesman Sean McCormack elaborated that "India is now taking a different place on the global stage, in terms of diplomacy, politics, and economy. . . . Aid programs have not caught up with these evolving realities."

[142] *Financial Times*, FT.com, December 6, 2006.

the CDB, tripled its lending to developing countries in five years, to over $15 billion in 2006, passing the U.S. Eximbank at $14 billion.

China is thus rising toward becoming the number one bilateral aid donor, with close linkages between long-term concessionary aid financing, commercially based Exim lending, and Chinese investment and exports. This development raises a number of policy issues that the MDBs and bilateral donors, within the OECD Development Assistance Committee (DAC), have been discussing and coordinating for decades. They include the following:

Transparency. Almost all detail about the CDB and China Exim programs remains secret, while full disclosure has been the foundation for aid donor coordination within the MDBs and among bilateral donors, including coordination in the field.

Financial terms. Little is known about the financial terms of Chinese aid. A minimum 25 percent grant element is required to qualify as ODA, but for poor countries, such as those in Africa, far softer terms are widely provided. It is not even known whether Chinese "aid" projects meet the 25 percent minimum grant element.

Tied aid. A major dispute among bilateral donors during the 1980s was over tying aid to donor country exports, and the resulting tied aid agreements in the DAC greatly limit such tying. A particular prohibition is placed on blending aid with Eximbank loans, a blending that can result in overall project lending at less than the 25 percent minimum grant element, and that can thus make the loan a prohibited export subsidy in terms of the WTO subsidy agreement.

Consistency with recipient country development strategy. This is a central subject of donor coordination, particularly when the recipients are the poorer countries most dependent on aid. Excessive aid loans at relatively high interest rates may cumulatively lead to an unsustainable debt buildup and default. Lower interest rates and grant assistance need to be coordinated, and priorities need to be placed on the most urgent and productive projects for achieving sustainable growth.

Policy conditions on the recipient country. This subject is wide ranging and controversial, going from economic policy conditions to ensure that the aid improves economic performance to good governance conditions related, among other things, to corruption, human rights violations, and excessive military expenditures.

From what is known about the Chinese aid program, all of these issues pose actual or potential problems. The most attention has been given to policy conditionality related to corruption, human rights, and charges of genocide in Darfur, in situations where China continues to provide financial assistance while other donors hold back. The president of the European Investment Bank, Philippe Maystadt, commented simply that Chinese banks in Asia and Africa, "don't bother about social or human rights conditions."[143] Concern has also been expressed about excessive new Chinese lending to countries that have recently defaulted and had their debt forgiven, and who are normally subject to some disciplines in resuming large foreign borrowing.

There is no question that CDB lending is closely linked to Chinese Export-Import Bank lending in support of Chinese investment and exports. CDB Governor Chen Yuan stated flatly, "We follow the biggest market players in China [abroad]."[144] CDB has extended credit to advanced technology companies Huawei and Lenovo, auto company Chery, and resource groups Chinese National Petroleum Corporation, Sinopec, and Minmetals. A blatant blend of export subsidy and foreign policy motivation, with adverse effect on the United States, is the $4 billion of CDB funding offered to Venezuela to build housing, roads, railways, and telecommunications. Aside from eliminating potential U.S. exporters from these projects, this Chinese funding will provide Venezuelan President Hugo Chavez an additional $4 billion to finance anti-American activities and programs throughout the Western Hemisphere. In return, China will presumably receive oil supply commitments.

The MDBs and the bilateral donors have their work cut out for them, and the principal proposal here is that collaboration rather than conflict between the Chinese and other aid programs be given a top priority for action. China has begun dialogue with other donors but still acts almost entirely on its own with a minimum of transparency, and the donor coordination process has a long way to go.

As for a more targeted U.S. response to burgeoning Chinese tied aid practices, one step could be what was done in the 1980s in response to Japanese and other donor tied aid, an action that helped bring about DAC agreements to curtail the practice.[145] The U.S. Eximbank was provided "War Chest" funding authority to match the credit terms of foreign tied aid loans for certain projects where U.S. exports were placed at a competitive disadvantage. For China, under current circumstances this program would not apply in some African and other poorest

---

[143] *Ibid.*

[144] *Ibid.*

[145] The War Chest experience is recounted in Ernest H. Preeg, *The Tied Aid Credit Issue: U.S. Export Competitiveness in Developing Countries* (CSIS, 1989). U.S. policy was summarized in a letter from Secretary of the Treasury Nicholas Brady and Export-Import Bank President John Macomber to House Speaker Thomas Foley: "The Administration recommends that the central thrust of the U.S. response to the tied aid credit problem should be vigorous new negotiations aimed at substantially reducing the commercial disadvantages for American exporters engendered by the tied aid credit practices of other countries. We also recommend that available budgetary resources be used aggressively to support these negotiations," p. 40.

countries where the U.S. Eximbank does not operate, but there could be Chinese projects in Latin America and Asia where U.S. Exim shots across the Chinese tied aid credit bow would send a useful message while maintaining a level financial playing field for U.S. exports. The "War Chest" title for this authority itself conveys a clear trade policy message.

Energy and the Environment

These vital and deeply integrated areas of policy have a very high commonality of interests and objectives for the United States, China, and India. They are also wide ranging into the fields of trade, investment, and new technology development. The challenges to closer collaboration, however, begin at home, at the national level, where strategies for achieving more secure energy supplies and a cleaner environment are ill-defined in all three countries, especially in terms of specific, adequately financed implementation. Consequently, the likely policy relationships ahead are similarly ill-defined.

For energy, there are the overlapping challenges of ensuring a secure oil supply at reasonable prices over at least the coming 10 to 20 years, making the longer term transition to renewable sources of energy, and, on the demand side, inducing less energy-intensive consumption, related most importantly to the automobile. For the environment, national levels of air, water, and land pollution feed into global concerns, especially about CO_2 emissions related to global warming, with China and the United States the number one and number two emitters and India destined to become number three.

The central policy question involves how the three emerging advanced technology superstates—and thus eventually the largest consumers of energy and the largest industrial polluters—can work more collaboratively to produce tangible results in meeting these challenges, which at the same time would work to build more positive overall political relationships. The question is fully understood by all three governments, but the answer remains elusive. A plethora of joint committees and technical groups are engaged, especially by the United States with each of the other two countries, but the results so far have been very limited. A U.S.-India Agreement on Environmental Cooperation and a United States-India Energy Security Cooperation Act are before the Congress, but little yet has happened in terms of funded, quantifiable goals. A report on U.S.-China Energy Cooperation refers to a U.S.-China Energy Policy Dialogue, the U.S.-China Oil and Gas Industry Forum, and the U.S.-China Peaceful Uses of Nuclear Technology Act, but specific projects are still mostly at the discussion stage. Likewise for a report on U.S.-China Environment, Science & Technology and Health Cooperation.

The time is ripe to select a few top-priority projects that would achieve demonstrable results and to make significant commitments on both sides. An analogy is the U.S.-Russia joint space program as a high visibility program of post-Cold War bilateral cooperation. The one big, high visibility initiative under way between the United States and India is the civilian nuclear energy agreement, which, if approved, should lead to deeply engaged private and public sector collaboration to produce civilian nuclear energy in India in the safest possible way. A similar mega initiative could be launched between the United States and China, such as in the area of clean coal technology. Binational research laboratories, with public and private sector participation, could be established with India and China for designated major, leading-edge technologies in the energy and environmental fields.

Looking ahead, two major developments deserve priority attention. The first is the post-Kyoto strategy for reducing CO_2 emissions. The United States, China, and India will have to be participants this time, as they were not for the Kyoto Agreement. If the strategy is not to commit to percentage reduction targets, which, in any event, would have little credibility under current circumstances in China and India, other more practical and credible actions and benchmarks will have to be on the table. It is not too soon to discuss a common approach.

The other major and more immediately threatening development is the likelihood of a major environmental crisis within China, and ways in which the United States could respond constructively. As to the scope of the threat, 16 of the 20 most polluted cities in the world are in China, 80 percent of the water in the East China Sea is polluted, and 10 percent of Chinese soil is toxic. Environmental laws exist in China, but are not implemented because responsibility rests with poorly funded local governments. There are reports of higher rates of infant mortality, cancer, and other health problems related to pollution, and public protests over pollution run into the thousands each year.[146] It is not clear where or when environmental crisis flashpoints will occur in China, but sooner rather than later a crisis will occur, and it is often in such times of crisis that international cooperation can produce positive results, in broader political as well as directly technical terms.

Even better, a couple of the most threatening situations can be identified now, before a crisis arises, and targeted for a high visibility

---

[146] These facts are from a presentation by Elizabeth Economy at the Chinese Studies Center in Washington on June 19, 2007. Much of the material is also contained in the statement she made before the U.S.-China Economic and Security Review Commission on June 14, 2007.

U.S.-China initiative. A similar preventive initiative would be welcome in India, where political collaboration would be easier to mobilize.

This concluding section of an otherwise action-oriented chapter can appear disappointing in its lack of specificity as to a U.S. policy response. This should in no way be taken to mean the subject is less important. To the contrary, energy and the environment are highly important and in need of a more creative and action-oriented policy response. Much of the problem in the United States, however, is the lack of clear direction and sense of priority for energy and environmental policies within the nation's political leadership, and a particular dearth of bipartisan purpose. In other words, before the international dimension can be fleshed out, there needs to be a forward-looking domestic policy agenda.

Annex 9

The International Trading System: From Here to Free Trade

The multilateral trading system, over 60 years, has been an extraordinary success in reducing barriers to trade and generating economic gains from increased trade. The system, however, has drifted into disarray, as the continuing process of trade liberalization and free trade has shifted away from the multilateral World Trade Organization (WTO) center to bilateral and regional agreements. The assimilation of the highly disparate developing country grouping into the system, which only began in earnest in the 1990s, has also been fraught with political discord, particularly within the WTO. This systemic disarray reduces the gains from trade and poses a threat to the overall rules-based trading system. The multilateral center needs to be reconsolidated through a decisive move to free trade on a multilateral, most-favored-nation basis.

Multilateral reconsolidation may appear unrealistic in view of the constant protectionist clatter in the United States about the outsourcing of jobs and unfair trading practices abroad, and of the six-year stalemate over limited tariff reductions in the WTO Doha Round. But the goal is not as daunting as it sounds. The U.S. trade deficit and related job loss are principally the result of misaligned currencies within the international financial system and, to a lesser degree, bilateral trade problems under negotiation, particularly with China. Meanwhile, trade barriers continue to come down through FTAs and unilaterally. Many nations have gone most of the way to free trade over the past 20 or more years, and the question about the future path can be posed as "why not go the rest of the way to multilateral free trade?"

Another reason for placing high priority on a reconsolidation and strengthening of the WTO trading system is that it is emerging as *primus inter pares*, by far, among the three organizational pillars of the Bretton Woods international economic system. The IMF role is declining as IMF loans dwindle and major trading nations move toward market-based floating exchange rates, albeit with a major immediate challenge on latter count. The World Bank role is also declining in relative terms, as economic aid becomes more concentrated in the poorest, least-developed grouping of countries. This is an extremely important challenge for alleviating poverty and other problems in these poorest countries, but it is narrowly targeted, and the World Bank is no longer at the global policy center in terms of trade, investment, and finance. The role of the WTO trading system, in contrast, is growing in absolute as well as relative terms. Trade as a share of GDP rises steadily, while becoming more deeply integrated with investment, technology, the services sector, and other areas of policy formerly considered domestic affairs. A look ahead shows that the deepening integration of trade with domestic economies will continue and that the current drift and uncertainty about the systemic underpinning for trade need to be addressed in forceful and forward-looking terms.

Yet another reason for addressing the revitalization of the WTO multilateral trading system here is that what will happen, for better or for worse, will depend, decisively, on the policy courses chosen by the three central protagonists of this study: the United States, China, and India. If the United States turns more inward and abandons its long-standing trade leadership role, the trading system will continue to drift apart. And if China and India adopt a minimalist approach to commitments within the WTO, little forward movement can be expected for the multilateral center of the system. The fate of the alternative course proposed here, that is a transition to multilateral free trade, will likewise depend on what evolves from the many forces in play in China and India, as recounted in the foregoing chapters.

This annex seeks to bring together these various strands of analysis to make the case that multilateral free trade is both feasible and desirable, and is in three parts. The first part addresses the question of why multilateral free trade, both in terms of the economic gains to be obtained and the conditions needed for assuring that these gains are equitably shared. The second part provides a brief analytic history of the trading system, as background for understanding the current circumstances and the prospects for change. Drawing on these two presentations, the third part proposes a policy path, centered on a two-stage multilateral free trade objective.

The Unprecedented Gains From Trade

The mutual gains from trade, based on comparative advantage, whereby nations specialize in producing goods and services for which they have a relative cost advantage, and then exchange or trade them to mutual advantage, have been largely obscured in recent public debate. In the United States, job losses related to the huge trade deficit tend to dominate, while at the multilateral level, within the WTO, a mercantilist orientation prevails, whereby lowering import barriers is an undesirable "concession." The political rhetoric in the United States is symbolized by the substitution of the loaded term "outsourcing of jobs" for "imports," while avoiding the counterpart "insourcing of jobs" term for exports. Aside from the negative connotation, these substitute terms are misleading, as explained below, and should be dropped. Exports and imports constitute the cross border exchange of goods and services that result in mutual gains from trade, which are currently at record levels.

Exactly how large the mutual gains from free trade would be is a complicated question with no precise answer. One rough estimate is a global gain in economic output of $2 trillion, as of 2002, amounting to about 3 percent of GDP for the United States and 6 percent for global GDP, with the largest relative gains by the newly industrialized grouping, prominently including China and India.[147] The analytic line of argument from which this estimate was made, and which points to even larger gains as of 2008 and beyond, has four components:

Trade is at a record high and a rising share of GDP.—Whatever the percentage gains from trade from producing at relatively lower cost in the exporting country, the absolute gains are now without precedent as trade rises to record shares of GDP. Global imports of merchandise as a share of global GDP rose from 8 percent in 1960 to 18 percent in 1980 to 27 percent in 2006. Comparable figures for rapidly growing trade in services would push these figures about 20 percent higher.

The "dynamic" gains from trade far outweigh the "static" gains.—Unfortunately, economic models for estimating the gains from trade, such as the computable general equilibrium (CGE) model, are limited principally to measuring the so-called "static" gains from trade, based on existing means of production. Trade growth, however, comes predominantly from the "dynamic" gains from trade generated by new investment and applied new technologies. These dynamic gains can be two to three times larger than the static gains, but they cannot be measured precisely and are therefore often ignored.

The gains from trade are increasing further as a result of cross border supply chain management.—The deepening process of international supply chain management, made possible by new information technologies, essentially involves subjecting each step in the production process to comparative cost analysis and locating each production step in the country with the lowest comparative costs.

Multilateral free trade provides a one-time bonus.—The move to across-the-board free trade—bilaterally and, even more so, multilaterally—provides a substantial extra one-time boost to increased trade and consequent economic gains. Free trade provides far greater political assurance that a free and open market will be maintained, which can facilitate trade-related, long-term investment and job creation. Moreover, the move to multilateral free trade also eliminates trade-distorting preferential tariffs from the existing skein of FTAs and the

---

[147] See Ernest H. Preeg, *From Here to Free Trade in Manufactures: Why and How* (Manufacturers Alliance/MAPI, 2003), Chapter 4, "The Gains from Free Trade and Related Investment," pp. 41-65. The $2 trillion estimate is on p. 48.

costly processing of "rules of origin" to determine if a product qualifies as originating within the FTA. More broadly, free trade reduces or eliminates the costs and delays of processing through customs. Particularly in many developing countries, such processing can be highly vulnerable to corruption.

This is the basic explanation of why the economic gains from multilateral free trade would be very large. There are also, however, conditions to be met to ensure that the gains are reasonably balanced and mutual among the participants. For the United States, four such conditions are significant, two of which are international in policy orientation and two domestic.

An international financial system of market-based exchange rates.— This was the subject of Chapter 8, although it cannot be overemphasized that the current misalignment of exchange rates through currency manipulation is a major threat to U.S. export competitiveness, particularly in the manufacturing sector, and to continued U.S. leadership for technological innovation. The resulting very large U.S. trade deficit is also the principal cause of the sharp decline in U.S. support for liberal trade on the part of both the Congress and the public, a decline that poses a major threat to a continued liberal trade policy of any sort.

Bilateral-track negotiations to assure compliance with WTO commitments.—This, too, was discussed earlier and in detail for China, in particular. There is no inconsistency, however, in pursuing multilateral free trade to eliminate border restrictions while continuing forceful bilateral negotiations on other trade-related issues. In fact, once free trade objectives are engaged, even at a preliminary stage, bilateral issues can be put in a more positive negotiating context.

Adjustment assistance for displaced workers in import-competing industries.—From a domestic perspective, growth in trade involves adjustment costs, including some loss of jobs in import-competing industries. The policy response is adjustment assistance to help such workers find new jobs. The United States has had a trade adjustment assistance program for almost 50 years, and it could be improved, particularly through orienting the payments more toward retraining and relocation, rather than being an extension of normal unemployment payments. In any event, the number of import-displaced workers is relatively small, especially for a growing economy where import growth can act to reduce job growth in a particular sector rather than to displace existing workers. Within a U.S. labor force of over 100 million, about 3 million people lose their jobs involuntarily each year, but only about 5 percent of the losses are trade related.

A domestic policy agenda in support of U.S. export competitiveness and continued leadership in technological innovation.—This is a critical challenge for the United States and is the subject of Chapter 10. Chinese and Indian domestic economic strategies, in particular, are heavily oriented to an increasingly competitive, knowledge-based global economy, while U.S. economic strategy, in some important respects, is moving in the wrong direction.

This, in brief, is the case explaining why and how free trade will bring large economic gains of mutual benefit. The international trading system has long been working in the direction of this goal, but the journey has been uneven and problem-laden.

A Brief History of the Multilateral Trading System

The post-war multilateral trading system was launched in 1947 through the General Agreement on Tariffs and Trade (GATT), which had two basic components. The first was a rules-based system for international trade that is subject to dispute settlement procedures, and the second was a series of multilateral trade negotiations to reduce historically high trade barriers on a reciprocal basis. The GATT was an extraordinary success in most respects, particularly in reducing trade barriers through eight "rounds" of negotiations, culminating with the Uruguay Round agreement of 1994.[148]

Until the Uruguay Round, however, GATT participation was essentially limited to the industrialized country grouping—Western Europe, the United States, Canada, Japan, Australia, and New Zealand. Industrial tariffs by this grouping were reduced from an average of over 15 percent in 1947 to less than 4 percent in 2006, thus going three quarters of the way to free trade. Some developing countries were GATT members, but they made relatively small reductions in their import barriers, while most developing countries and almost all of the Soviet bloc were not members. Trade liberalization, moreover, was limited principally to industrial tariffs, while barriers to agricultural trade, despite strenuous efforts, remained very high. It was also noteworthy that the United States always played a strong leadership

[148] The two most important rounds were the Kennedy Round of the 1960s, which centered on the U.S.-EU relationship and sought to reduce tariffs by half so as to avoid an inward-directed European free trade zone, and the Uruguay Round as discussed here. For analytic histories of these two rounds, see Ernest H. Preeg, *Traders and Diplomats: An Analysis of the Kennedy Round of Negotiations Under the General Agreement on Tariffs and Trade* (The Brookings Institution, 1970), and Ernest H. Preeg, *Traders in a Brave New World: The Uruguay Round and the Future of the International Trading System* (University of Chicago Press, 1995). The author participated in both Rounds.

role for trade liberalization, proposing the most ambitious objectives at the outset of each round, which led to reasonably large final results.

This all changed greatly with the Uruguay Round agreement. The mandate of the trading system was extended to cover trade in services, some trade-related investment measures, and intellectual property rights. Agriculture was brought fully within the system through general commitments, such as to use tariffs rather than quota restrictions for imports and to phase down export subsidies, but protection remained extremely high. In organizational terms, the World Trade Organization (WTO) was created which incorporated the GATT and greatly strengthened dispute settlement procedures, in particular. Participation was also broadened to include, more fully, developing countries, which had to reduce their import barriers and adopt a broad range of WTO commitments to become members in the new organization. In parallel, many more developing and former Soviet bloc countries applied for membership, most importantly China. GATT membership rose from 99 in 1990 to 151 in 2007.

There was great hope that the WTO would emerge as a truly multilateral system of trade, dedicated to a further substantial reduction in trade barriers, with an even broader mandate, perhaps including international investment. This optimism quickly faded, however, with the blame for lack of progress widely shared. Initial attempts to organize a new trade round failed at the 1999 WTO ministerial meeting in Seattle, when the United States alone insisted on bringing international labor standards into the WTO system, while the EU and Japan resisted commitments in agricultural trade, and developing countries were divided and reluctant to make any specific commitments.[149] The Doha Round was launched in November 2001, in the aftermath of the attack on September 11, with a strong sense of international political solidarity but with even greater reluctance by the developing country grouping to reduce its trade barriers, in what was officially named "The Doha Development Agenda." The Doha Round then settled into a six-year impasse over the basic modalities, or formulas, for reducing import barriers, which in past rounds had been agreed to during the first year or two, after which the real negotiations began, principally over limiting exceptions to the formulas. For the presentation here, geared to assessing the prospects for moving to multilateral free trade, three key issues of the Doha Round impasse are assessed, in each case with

[149] For an analytic account of the Seattle meeting, see Ernest H. Preeg, *Charting a Course for the Multilateral Trading System: The Seattle Ministerial Meeting and Beyond* (The Group of Thirty, 1999). The author was an accredited journalist to this meeting.

invidious comparison to successful FTAs, where these three issues are not a problem, either by definition or by design.

The Primacy of Agriculture

Major liberalization of agricultural trade has been the *sine qua non* for the Doha Round. The failed ministerial meeting in Cancun, Mexico, in 2002, was devoted to five days of almost exclusive debate over agriculture, while the other two areas of market access negotiations, trade in nonagricultural products and services, were never seriously addressed by ministers, despite the fact that trade in these sectors is 10 times larger than that in agriculture. Agricultural results were always limited in previous GATT negotiations, and the breakthrough in the Uruguay Round was the result of a confluence of unique political circumstances between the United States and the EU.[150] The fact is that there is an extraordinary political divide in trade policy between agricultural and nonagricultural trade in almost all major trading nations—highly protectionist in the former and tending strongly toward free trade in the latter. Agricultural protection, in one form or another, rides high in the United States, the EU, Japan, South Korea, and among almost all developing countries, including China and India. This protectionism does not mean that some liberalization of agricultural trade is not possible, but it will remain moderate in degree until the deeply embedded political commitment to protect farmers in major trading nations weakens. In this context, making major trade liberalization for agriculture the make-or-break issue in the Doha Round has resulted in a broken round.

In FTAs, in contrast, agriculture is generally included, but with more modest objectives and results. Selected tariffs are reduced or eliminated and some marketing restrictions relaxed, but when a major trade barrier is confronted, as, for example, Korean rice imports related to the U.S.-Korea FTA, it is put aside. Likewise, principal U.S. agricultural protection is in the form of domestic subsidies rather than border restrictions, which are not feasible for reduction within bilateral FTAs, and are thus conveniently excluded, which greatly reduces the common denominator of results in the agricultural sector.

More fruitful results for agriculture may be possible in the context of multilateral free trade for nonagricultural trade, as described in the following section, but the agricultural cart should not be placed before the nonagricultural horse, as happened in the Doha Round.

---

[150] See Preeg, *op. cit.*, 1995, related to the Blair House Accord, pp. 143-147.

The Mercantilist Mindset

GATT rounds of negotiation were based on "reciprocity," that is comparable reductions of import barriers by all participants. In practice, this led to tariff reductions on imports being termed "concessions," which needed to be offset by reciprocal "benefits" from market access for exports. This approach had an implied mercantilist orientation, whereby imports were bad and exports were good. Within the setting of reciprocal reductions by all participants, this conceptual problem was occasionally discussed and rationalized as a political necessity in order to obtain political support for the final agreement, whereby the adverse impact on domestic industries faced with increased import competition could be justified in terms of increased benefits for export industries.

This mercantilist rationale, however, became far more prominent, if not overriding, in the Doha Round, particularly by developing countries, which implied that a reduction in import barriers conflicted with their development strategies. Discussion of mutual benefits from mutual reduction in import barriers—the free trade rationale—essentially disappeared from Doha Round deliberations.

Again, the contrast with FTAs is stark, in effect by definition. In FTAs, the two parties agree at the outset that there is a mutual interest in eliminating import barriers in both directions. Developing countries, moreover, generally starting with much higher import barriers, agree in their own self-interest to much larger absolute reductions in their import barriers. This leads to the third issue of nonreciprocity for developing countries.

Nonreciprocity for Developing Countries

"Special and differential" treatment for developing countries within the GATT dates back to the 1960s, and is anchored in the provision that developing countries do not have to provide reciprocity in reducing trade barriers. This was not a significant issue until some developing countries, particularly in East Asia, became large exporters of manufactures in the 1970s, when a revised commitment was adopted for developing countries to "participate more fully . . . with the progressive development of their economies and improvement in their trade situation."[151] In other words, there would be a transition toward full reciprocity by the more export-competitive developing countries.

---

[151] This is the Enabling Clause adopted at the conclusion of the Tokyo Round in 1979.

The United States and the EU were even more concerned about fuller reciprocity in the Uruguay Round. The earlier language about fuller participation as development progressed was restated, and the more advanced developing countries were pressed to provide broader market access commitments, which they did. The United States established an interagency group to develop a strategy for developing countries in the Uruguay Round, focused on upward movement toward graduation and full reciprocity, beginning with the four East Asian tigers—Hong Kong, Singapore, South Korea, and Taiwan. The protectionist approach to trade negotiations was condemned in a speech by Acting Secretary of the Treasury Peter McPherson, who had previously served for six years as Administrator of the U.S. Agency for International Development. The protectionist approach, he stated, is "built on false assumptions," and the special and differential provisions of the GATT, including nonreciprocity, "not only encouraged bad development policies, they were used for cover for protectionism that has nothing to do with development."[152]

This strong stand for increased reciprocity by the more advanced developing countries, as being in their own development self-interest, was lost in the Doha Round mandate, which provides blanket non-reciprocity for all developing countries and has been the foundation for far less reduction in their trade barriers, as defined in the proposed formulas for tariff reductions, with widely different coefficients for the developed and developing country groupings. Since the 1994 Uruguay agreement, moreover, a number of newly industrialized export power-houses, including China and, more recently, India, have pulled far ahead of the majority of developing countries, which makes the blanket nonreciprocity formulas, with the same coefficients for all developing countries, far more difficult if not impossible to justify in commercial, as well as developmental terms. In effect, the same nonreciprocity formula applies to China as to Chad. This fundamental North/South divide over reciprocity, together with the mercantilist mindset described above, has constituted a fatal flaw to a successful outcome for the Doha Round.

And once again, FTAs, in contrast, are successfully concluded, often within a year or two, while excluding, by definition, the issue of

[152] For a discussion of the nonreciprocity issue in the Uruguay Round, including the quotes here, see Preeg, *op. cit.*, 1995, pp. 70-74. The author chaired the interagency group while serving as chief economist at the U.S. Agency for International Development, and McPherson's speech was the final product.

nonreciprocity. Each participant simply goes to zero tariffs on a fully reciprocal basis. Some special treatment has been provided to developing country participants, but it is principally technical assistance and other aid projects to help the country adapt to free trade and reap the maximum economic benefits.

These three issues—agriculture, the mercantilist mindset, and nonreciprocity for developing countries—have created sharply different political settings for what has been happening in the Doha Round and in FTA negotiations since 2001. They also largely define a crossroads for alternative paths as to what happens next—a continued impasse over reciprocity or the quest for a multilateral free trade agreement. Unsurprisingly, based on the three issues recounted, the FTA path is recommended here.

From Here to Multilateral Free Trade: A Proposal

The multilateral trading system is in disarray, but it can be reconsolidated and strengthened by one simple yet bold initiative: a multilateral free trade agreement for the nonagricultural sector, which accounts for over 90 percent of merchandise trade.

For 60 years, the world trading system has been moving toward free trade. The Kennedy Round of the 1960s alone reduced tariffs by 50 percent across the board, with limited exceptions. Surprisingly, however, there has never been serious discussion within the GATT or the WTO as to what the ultimate objective for trade liberalization should be and whether, in particular, it should be free trade. In earlier decades, EU leaders avoided the subject because they believed that a high visibility "common external tariff" was necessary to maintain political cohesion. In recent years, nonreciprocity for developing countries has centered on how to do less rather than more tariff reductions.

Outside the WTO, after the Uruguay Round, however, interest began to build to move the final step to multilateral free trade. Fred Bergsten, the Director of the Peterson Institute for International Economics in Washington, called for a "Grand Bargain," so as to seize the "enormous opportunities for further economic gain in eliminating remaining tariffs and nontariff barriers."[153] A book of essays on post-Uruguay Round trade strategy was entitled *From Here to Free*

[153] See C. Fred Bergsten, "Globalizing Free Trade: The Ascent of Regionalism," *Foreign Affairs* (May/June 1996), pp. 105-120.

Trade.[154] Most importantly, governments were moving to free trade bilaterally and regionally in Europe, the Western Hemisphere, East Asia, and across the Pacific. The logical next step would be for some of these governments to redirect the free trade momentum back into the WTO on a multilateral basis.

The United States finally rose to the occasion in November 2002, in the Doha Round, with its initial proposal for multilateral free trade in the nonagricultural sector. The proposal was received positively by some participants, including Singapore, Hong Kong, Taiwan, Australia, New Zealand, and Uruguay. Other major participants, however, were negative. The EU and Japan considered the proposal impracticable and unrealistic, while the Indian representative called it "clearly unfair."[155] The U.S. proposal was never discussed seriously, and it was quietly shunted aside as negotiators concentrated on agricultural trade and nonreciprocal tariff-cutting formulas of far more modest intent. The free trade objective within the Doha Round nevertheless continued to be pursued on a sectoral basis, again with U.S. leadership, and again with no progress beyond sectoral discussion groups.

This is the starting point for the proposal presented here. Multilateral free trade by sector, and for the entire nonagricultural sector, have been on the Doha Round negotiating table for five years, and it now deserves to be addressed seriously with a view toward future negotiation, especially in view of the disappointing experience of the Doha Round and the FTAs achieved in all major regions. The first step is to agree on the objective, which then can best be implemented on a two-stage basis.

Establishing the Multilateral Free Trade Objective

Agreement on a post-Doha Round objective of multilateral free trade for the nonagricultural sector should be pursued on both the Asia-Pacific Economic Cooperation (APEC) and the WTO fronts. APEC established free trade within the region as the objective at a summit

---

[154] See Ernest H. Preeg, *From Here to Free Trade: Essays in Post-Uruguay Trade Strategy* (University of Chicago Press and CSIS, 1996). The book's six essays included discussion of free trade by region, and concluded, in the final essay, that "a WTO free trade Grand Bargain does not appear politically feasible at this time, although perhaps it will become so in ten years' time as the trading system evolves—and hopefully converges—further at both the multilateral and regional free trade levels" (p. 120). The ten years are now up, and sufficient evolution and convergence have occurred that now is the time to act.

[155] See Preeg, *op. cit.*, 2003, p. 3.

meeting in 1994, with full implementation by 2020. This objective should be reaffirmed and elaborated. The specific path to free trade was never agreed and, in view of the recent proliferation of FTAs within the region, an FTA "building block" approach should be endorsed, which would encourage further FTAs within Asia and across the Pacific. Another key element of the APEC free trade objective was that it would be "open-ended," that is, open to countries outside the region who wished to join. In earlier years, this issue was discussed largely in terms of possible EU engagement, but now India is an even more pressing candidate from within Asia and should be invited to join the APEC commitment to regional free trade.

The initiative to inscribe a multilateral free trade agreement in the WTO work program, with a view to possible future negotiation, would build from the APEC momentum. If APEC members, including China and India, were to support such a proposal in the WTO as an extension of the open-ended provision of the APEC free trade commitment, the proposal would surely be adopted. Far better, however, would be to coordinate such an initiative beforehand, among APEC, the EU, and Brazil, in particular. Such a joint approach would make the work program a truly multilateral initiative. The specific guidelines for the work program, such as the relationship of this objective to other areas of the WTO work program, and the plurilateral trade threshold concept, are addressed below.

Once the multilateral free trade objective is reaffirmed within APEC and inscribed in the WTO work program, the way would be open to further FTA building blocks which, in turn, would build more momentum toward the ultimate step of consolidation within a multilateral free trade agreement.

Implementation Stage One: Further FTA Building Blocks

Although additional FTAs could be negotiated in all regions, the Asia-Pacific region would be the most important by far. Chinese, Japanese, and South Korean FTAs with ASEAN, already in place for China and under negotiation for the other two, would increase the incentives for an ASEAN plus three agreement. Further U.S. FTAs across the Pacific with Thailand and Malaysia, where negotiations had begun before U.S. Trade Promotion Authority negotiating authority expired in June 2007, and with Vietnam, Taiwan, and, most importantly, Japan would confirm that the regional free trade objective is trans-Pacific, including the United States, and would thus end discussion of an East Asia trading bloc which excludes the United States. A U.S.-Taiwan FTA should not be a major political problem for China because

Taiwan is a member of APEC and the WTO. Even better, Taiwan, following the path of Singapore and Hong Kong, could move unilaterally to free trade and thus give an important leading-edge, free trade-impulse from the center of the Asia-Pacific region, for others, including China, to follow.

India could play a critical role in the overall FTA building block process. FTA negotiations already under way with ASEAN and the EU make India a strategic bridge between APEC and Europe. Further Indian FTA initiatives with Japan, South Korea, and, most importantly, the United States, would make the ultimate step of multilateral free trade almost inevitable.

The question of timing for moving from this building block stage one to the multilateral negotiating stage two cannot be projected with precision. A thorough review within the WTO work program would probably take a couple of years. The biggest question, however, concerns China. The surging Chinese trade surplus in manufactures, at $440 billion in 2007 and continuing to grow, makes many nations, including in Europe, Latin America, and, of course, the United States, extremely reluctant to consider opening their markets freely to imports from China. This situation derives principally from the Chinese currency misalignment and industrial policy problems discussed earlier, and major progress would have to be made in these areas before others would consider free trade with China. For the United States and China, there is also the underlying political issue of whether they are competing, if not adversarial, global powers, or are economic partners within a projected system of global free trade. As a rough estimate and assuming progress on the issues described here, stage two could be reached in three to five years, although this assumption is highly uncertain.

Implementation Stage Two: A Multilateral Free Trade Agreement

A detailed analysis and recommendations for negotiating a multilateral free trade agreement is contained in *From Here to Free Trade in Manufactures: Why and How.*[156] That work elaborated the U.S. 2002 free trade proposal in the Doha Round, was circulated to the principal delegates and senior WTO secretariat before the September 2003 ministerial meeting in Cancun, and was discussed there with delegates by the author, but regrettably to no significant effect. The presentation here is limited to three basic points, with other details provided in the referenced study.

---

[156] Preeg, *op. cit.*, 2003, Chapters 7-9.

The nonagricultural sector scope of the agreement.—The non-agricultural goods sector is dominant in trade, accounting for over 90 percent of merchandise trade and about 75 percent of total trade including services. A free trade agreement for this sector would thus definitively move the trading system to multilateral free trade. Separate agricultural and nonagricultural negotiating groups have always prevailed in GATT rounds and now in the Doha Round because the systems of protection are fundamentally different. Domestic and export subsidies are widespread if not dominant in protecting agricultural production, while they are basically prohibited for nonagricultural trade. This is what made the U.S. free trade proposal limited to the nonagricultural negotiating group in the Doha Round far more feasible than free trade for all sectors. The nonagricultural scope should therefore be adopted for the multilateral free trade agreement, with agriculture and other areas of trade policy treated in the same timeframe, but separately, as addressed in the third point.

The 90 percent threshold.—The free trade agreement needs to include the advanced industrialized and newly industrialized groupings that account for over 90 percent of trade in manufactures, which is essentially what is at stake in terms of a free trade agreement because imports of petroleum and raw materials are already predominantly duty free. The most practical way of negotiating such an agreement is on a "plurilateral" basis, whereby participants would have to reach a threshold accounting for at least 90 percent of total nonagricultural imports. There is precedent for such a plurilateral free trade approach in the WTO in the information technology sector free trade agreement of 1996. This approach would permit a large number of lesser developed, mostly smaller countries to gain free access to the 90-plus percent of export markets of the more industrialized free trade participants, while permitting these lesser developed to proceed at their own pace toward free trade. The critical practical consequence of this plurilateral approach is that the actual negotiations would be by the 20 to 30 largest trading WTO participants, with the EU a single participant represented by the Commission, rather than by the entire 150 WTO members.

The connection with other WTO programs.—A post-Doha Round work program to assess a possible future free trade agreement for the nonagricultural sector would go forward in parallel with work programs for other sectors, including agriculture and trade in services. These other programs would be designed in terms of addressing possible future objectives in light of the Doha Round outcome. Agriculture would certainly need a basic reassessment. Trade in services might proceed better in terms of sectoral agreements, on an expanded plurilateral basis as in the WTO financial services and telecommunications agreements.

A possible new area for future negotiation is investment policy, which some proposed, unsuccessfully, for the Doha Round agenda. In view of basic differences in investment policy, and new issues such as the rapid growth of sovereign investment funds, the next step for investment might best be a revived OECD negotiation, open-ended for nonmembers to join. Other issues, certainly including competition policy related to the "trade remedy" of anti-dumping duties, would inevitably be part of a future WTO work program. The connection made here is that if a multilateral free trade objective were pursued for the dominant nonagricultural sector of trade, there should be positive spinoff effects for more ambitious trade liberalization elsewhere.

This is the Bergsten Grand Bargain, revisited and refined, a strategy to achieve multilateral free trade in nonagricultural trade over the coming five to ten years. Its development needs careful and serious discussion in various forums. APEC and the WTO have been proposed here as starting points. The venerable OECD trade committee would inevitably get into the act. Bilateral discussions and working groups would also be vital. For the United States, such discussion would be natural with the EU and Japan. Decisive U.S. bilateral discussion channels, however, would also be with China and India, within the Strategic Economic Dialogue with the former and in a suitable ministerial level economic framework with the latter. Multilateral free trade is clearly on the horizon, and it is time to focus the official minds on ways to achieve it.

CHAPTER 10

The Domestic Economic Policy Response

The two preceding chapters dealt with the international policy response to the Chinese and Indian challenges to U.S. technological leadership and export competitiveness. This chapter addresses the corresponding domestic policy agenda. The two responses should be mutually reinforcing, although they differ significantly in timing and political context.

As for timing, the immediate problems are principally international, especially the problem of currency misalignment, which has provided very large and unfair cost advantages to China and other currency manipulators and is thus responsible for much of the shift of R&D and export-oriented investment to these countries. The international financial and trade issues, moreover, can be dealt with more quickly, such as through exchange rate adjustment and elimination of targeted unfair trade restrictions. The domestic policy agenda, in contrast, is longer term in development and implementation, even in the best of circumstances, while at the same time of greater importance for the longer term outcome. Actions to address relative weakness in education and regulatory policies that discourage R&D and investment in the United States, for example, can take a number of years to formulate and implement, and yet they will be more important and decisive over the longer term, particularly as the current international problems are progressively dealt with.

As for political context, the objectives on the domestic policy agenda are more complex and controversial, often requiring large public expenditures, and almost always competing with other domestic interests. Again, by comparison, currency realignment is direct and unambiguous in terms of the U.S. objective, whereas the domestic issue of how, for example, to allocate publicly funded R&D between industrial technologies and health care objectives engages strong conflicting interests, with health care generally carrying the day.

The presentation here begins with a discussion of the need to build a sense of national purpose in responding to the international advanced technology challenge, as the point of departure for achieving substantial results on the domestic front, in particular. This is followed by an annotated listing of the principal domestic policy issues that should be addressed, but with considerably less detail than was provided for the international issues. The intent is more to lay out the framework of the issue for a broad-based policy response, and to highlight the international dimension of the domestic issues, which is often neglected. The chapter concludes with a commentary drawing on the cherished American philosopher Pogo, entitled "Our Own Worst Enemy."

Building a Sense of National Purpose

There is no sharper contrast between the economic strategies of China and India and that of the United States than the sense of national purpose and historic destiny embodied in the development of globally competitive advanced technology national economies. Chinese economic strategy statements, as recounted earlier, center on the objective of creating indigenous innovation and export-oriented, high-tech companies. The Indian message of building a knowledge-based, advanced technology society is ubiquitous, appearing even on lamppost signs throughout New Delhi. As quoted earlier, Chinese Premier Wen Jiabao summed up the joint Indo-Chinese vision as signifying the coming of the Asian century of the information technology industry. With this vision front and center, both governments are able to mobilize substantial financial resources and take other difficult steps to realize their national objective.

In the United States, in sad contrast, there are conflicting views and no broadly shared sense of national purpose or destiny. Influential voices claim that the large trade deficit in manufactures, including for advanced technology products, is not a problem or something to worry about, while many interest groups seek to regulate and tax American companies in ways that reduce rather than enhance their international competitiveness. Experts in business and academia speak out about the vital U.S. interest in maintaining technological leadership and export competitiveness, but they receive limited attention from political leaders and the general public. In 2004, the President's Council of Advisors on Science and Technology, comprised of 24 distinguished academic and business leaders, concluded:

> The loss of U.S. high-tech leadership would have serious detrimental effects on the Nation's economic security and the citizens' standard of living. While not in imminent jeopardy, a continuation of current trends could result in a breakdown in the web of "innovation eco-systems" that drive the successful U.S. innovation system. . . . In particular, the entry of China into the high-technology arena has created a new level of nervousness on the part of many industry and academic professionals. In part, this results from China's size and its commitment to a

> high-tech industrial policy. . . . China also has a flexible, entrepreneurial culture, which some of its neighbors do not. . . . China has an interest in seeing economic benefits accrue to Chinese companies rather than to foreign competitors. . . . It is expected that China's efforts to develop leading-edge high technology ecosystems will be significant, continue for a long time, and gain extensive assistance from foreign investment.[157]

Regrettably, the report received no official recognition and quickly receded into the depths of the Executive Office website. President Bush, in his 2006 State of the Union message, raised the issue: "In a dynamic world economy, we are seeing new competitors, like China and India, and this creates uncertainty. . . . We must continue to lead the world in human talent and creativity. . . . Tonight I announce an American Competitiveness Initiative to encourage innovation throughout the economy." Again, however, little happened as a result. Legislation was proposed for modest additional funding for education and public sector R&D, but no serious discussion or action took place in the Congress over the ensuing two years.

A broader vision and deeper sense of national purpose are clearly needed from the political leadership in both parties if effective remedial action is to be forthcoming. U.S. technological leadership and export competitiveness are under serious threat. The October 2005 report of the National Academies of Science, which triggered the President's competitiveness initiative three months later, made the point forcefully. The gathering Chinese technological strength was highlighted, for example by observing that of 120 chemical plants being built around the world with price tags of $1 billion or more, one is in the United States and 50 are in China. The report then concluded:

> Having reviewed trends in the United States and abroad, the Committee is deeply concerned that the scientific and technical building blocks of our economic leadership are eroding at a time when many other nations are gathering strength. . . . We fear the abruptness with which a lead in science and technology can be lost—and

[157] President's Council of Advisors on Science and Technology, *Sustaining the Nation's Innovation Ecosystems: Report on Information Technology Manufacturing and Competitiveness* (Executive Office of the President, 2004).

> the difficulty of recovering a lead once lost—if indeed it can be regained at all.[158]

This is the challenge which needs to be met with a strong sense of national purpose and commitment. Unfortunately, during the two parties' presidential debates of 2007, the subject was never raised as a matter of high national priority by any of the many candidates or by the media and other questioners. One can hope that it will be given more attention during the final course of the election campaign, and in the program of the new administration. This study is designed to give analytic support for such an awakening.

As for the domestic agenda, the issues discussed below require priority attention, which they seldom receive in terms of their impact on international competitiveness. One procedural step would be to require a statement of international economic impact before taking significant executive or legislative decisions. Such questions of international economic impact are highlighted in the issue commentary here.

The Domestic Agenda

The domestic economic policy agenda for strengthening U.S. technological leadership and export competitiveness is presented in terms of seven broad areas of policy and one example of a specific piece of legislation—the Sarbanes-Oxley law—which was adopted with virtually no consideration of the adverse international impact on the U.S. economy. The seven are not necessarily in priority order, except for education, which is clearly number one.

Education.—The relative weakness of U.S. education in science and technology-related skills is well-known, both at the K-12 and university levels. For K-12, math and science proficiency is poor by international comparison, with American eighth graders ranking 15th among students in 45 countries tested for math and 9th out of 45 for science.[159] School curricula are spread in other directions, many teachers are not trained in math and science, and relative performance tends to weaken progressively into the more decisive higher grades. Inner city schools record especially low performance.

---

[158] National Academy of Sciences, *Rising Above the Gathering Storm: Energizing and Employing America for a Brighter Economic Future* (Preliminary Report, October 2005), p. 3.

[159] American Electronics Association, *We Are Still Losing the Competitive Advantage: Now Is the Time to Act* (March 2007), p. 22.

One remedy would be a large-scale scholarship program for undergraduate majors in math and science who plan to be teachers. This remedy is, in fact, being pursued through the American Competitiveness Act of 2007. A broader response would be to provide more competition among schools for math and science performance, permitting parents to send their children to the more demanding and higher performing schools. This competition is happening in private schools in the United States, as it is in India and China, but such choice is readily affordable only by wealthy and upper-middle class families. Comparable choice for lower-income parents would require more charter schools and vouchers. There should be no question, however, that low-quality math and science education at the K-12 level puts low-achieving American students at a competitive disadvantage relative to students in higher quality schools at home and, increasingly, abroad.

At the university level, the United States has top-quality engineering schools and science programs, but the quantity of graduates has leveled off, while at the same time the share of foreign graduates is rising. The number of engineering undergraduate degrees grew only 2 percent from 1995 to 2004, while in the latter year the share of doctoral degrees awarded to foreign students was 50 percent for computer science, 53 percent for math, and 60 percent for engineering.[160] Moreover, as recounted in Chapter 4, an increasing number of foreign graduates are returning home to India and China.

This is an area where the terms of the public policy debate need to be in broader international terms. Current discussion about the adequacy of engineering graduates focuses on short-term domestic job opportunities at a time when the manufacturing sector suffers from a trade deficit of more than $600 billion, which equates to about 40 percent of value added in the domestic manufacturing sector. If the international economic strategy, however, calls for strengthening international competitiveness and reducing the trade deficit, perhaps to $200 billion, which is comparable in relative terms to what happened in the late 1980s, this would project to an increase of $400 billion in domestic manufacturing production and several million additional jobs, including tens of thousands of engineers. But would these engineers be forthcoming from a university system that has remained flat while the trade deficit soared? The inability of the university system to respond would realize the National Academies' warning of the difficulty of regaining a science and technology lead once lost.

A bold and high-visibility initiative which would help build a sense of national purpose along with professional skills would be a repeat of the National Defense Education Act (NDEA) of the early 1960s that

---

[160] *Ibid*, p. 23.

was aimed at meeting the challenge of the Soviet manned space program. The NDEA was a large program of Ph.D. fellowships in science, engineering, and other advanced technology studies. A companion initiative would be to grant a green card to all foreign students receiving graduate degrees in science, math, and engineering.

Public sector R&D funding.—This is a relatively small part of R&D spending, but it is the principal source of funding for basic research, which is of critical importance for innovation leading to later commercial development and application. Such basic research has long been a major strength for the United States, although in recent years, with the end of the Cold War, the U.S. commitment to non-defense federal funding of R&D has declined substantially, and most sharply for technology-oriented R&D administered by the National Science Foundation (NSF). Federal funding of R&D as a percent of GDP declined from 1.25 percent in 1985 to 0.80 percent in 2004. Within this total, the share of funding for technology-oriented R&D dropped from 46 percent to 32 percent, while that for life science and health care increased from 30 percent to 54 percent.[161]

In this critical area, U.S. international leadership is thus being squeezed from two directions. On the domestic side, health care has risen to primacy, while the international challenge to U.S. technological leadership today does not have a domestic constituency anywhere near what it had during the Cold War era, when support for civilian basic research was closely linked to defense needs. On the international side, in contrast, China has recently placed higher priority on basic research, through increased funding to universities and public research centers, with incentives for collaboration with private companies.

As for the U.S. policy response, a higher priority should be placed on technology-oriented R&D, particularly through the NSF and perhaps through other programs as well. This has begun to happen. Appropriations for the NSF increased from $3.9 billion in 2000 to $5.6 billion in 2004 and to $5.9 billion in 2007. In any event, the international challenge to U.S. technological leadership from China and others should be a major part of congressional hearings on NSF budget requests, which it has not been. There is also the question of relative priorities between technology innovation and health care research. Health care may be number one, but 54/32? Perhaps the rallying cry for those concerned about the international technological challenge should be "54/40 or fight."

Corporate tax rates.—Corporate tax rates vary widely among nations, and are quickly factored into corporate investment decisions. Other countries have lowered substantially their tax rates in recent

---

[161] *Ibid*, p. 18.

years compared with the United States, with a resulting decline in cost competitiveness for production in the United States, and an increased incentive to invest abroad. Within the OECD industrialized country grouping, from 1996 to 2006, 22 nations reduced their tax rates, 6, including the United States, were unchanged, and only 1 increased rates.[162] Japan reduced its rate from 50 percent to 39.5 percent, and is now in an approximate tie with the United States, at 39.3 percent, for the highest rate. The German rate was reduced from 58 percent to 36 percent, and is scheduled for a further cut to 30 percent in 2008. Italy reduced its rate from 50 percent to 37 percent, and Poland from 40 percent to 19 percent. Ireland had the lowest tax rate, at 13 percent, and was the one country showing a small increase since 1996. This relative tax disadvantage for production in the United States compared with other industrialized countries, moreover, can pale by comparison with tax incentives offered by China, India, and other newly industrialized nations, especially for new investment in advanced technology, export-oriented industries. The basic corporate rate in China is being reduced to 25 percent, but often with 5 to 10 year tax holidays for such industries.

Clearly, U.S. corporate tax rates can no longer be assessed only in their domestic context but need also to be compared on an international basis, as a significant factor in determining where investment and job creation will take place. This consideration is also increasingly recognized. Secretary of the Treasury Henry Paulson convened a U.S. business tax competitiveness conference in July 2007 and opened it with the question: "What is the impact of the business tax system on the competitiveness of U.S. businesses and how important are taxes relative to other factors which determine our economic competitiveness?" Participant Alan Greenspan responded that the relatively high U.S. tax rates are "undermining the competitiveness of American workers."[163]

The problem is thus clearly identified, but the outlook for bipartisan support for basic tax reform remains bleak. Tax reform is a highly controversial subject and extremely difficult to achieve, with action usually limited to amendments to existing tax codes, in the form of opening or closing loopholes. And yet basic tax reform is needed, both for domestic reasons to improve productivity and for restoring international competitiveness. Corporate tax reform should address the problem of dual taxation of American companies, on profits first and again on dividend and capital gains distributions, and the problem of

---

[162] OECD, *Making the Most of Globalisation*, Preliminary Edition (2007), p. 18. These are statutory rates that include state as well as federal taxes.
[163] Economist.com/finance, August 2, 2007.

the myriad of tax deductions which can have adverse impact on efficiency at home and competitiveness abroad. The broad objective should be a reduction in taxes through a single, lower rate, financed in part by the elimination of most if not all targeted deductions. Such reform would have the additional benefit of permitting thousands of the most innovative and over-achieving Americans to shift from being corporate tax lawyers and lobbyists to doing other, more productive work. In any event, official consideration of basic tax reform should include an international economic impact statement.

Tort reform.—Tort litigation in the United States has been growing at 10 percent per year for several decades, and more than 15 million lawsuits are now filed each year. The direct cost to U.S. firms from tort litigation is estimated to be about $250 billion per year. Additional indirect costs include expenditures by companies to avoid lawsuits and a general tendency to be risk-averse, which has an especially negative impact on technological innovation. The direct costs alone amount to 2 percent of GDP, compared with less than 1 percent in Japan, Canada, France, and the United Kingdom, and probably a much smaller share in China and India. The high financial cost and other international disadvantages to American companies are thus substantial, with technology-intensive industries especially hard hit because they tend to be more vulnerable to lawsuits.[164]

Tort reform is a broadly recognized objective, but progress is slow and uncertain in the face of strong resistance from the tort lawyers, who receive $40 billion per year in fees, a significant part of which is used for political campaign contributions and other means to achieve their objectives. The Class Action Fairness Act of 2005 transferred jurisdiction for multistate class action suits to federal courts, and has resulted in some rules of settlement more favorable to defendants. Medical malpractice tort reform is moving forward at the state level where jurisdiction centers, with legislation enacted in nine states in 2006. Broader reform is far less likely, however, particularly to alleviate the costs for U.S. multinational corporations, who lack the political support given to health care providers.

As with the other domestic issues discussed here, the question is how to put tort reform in a broader international context, one related to the comparative cost disadvantages to production and jobs in the United States, particularly for innovation and R&D expenditures. One

---

[164] A comprehensive assessment of the crisis in tort litigation, from which the figures cited here are drawn, is contained in Frederick T. Stocker, ed., *I Pay, You Pay, We All Pay: How the Growing Tort Crisis Undermines the U.S. Economy and the American System of Justice* (Manufacturers Alliance/MAPI, 2003).

way would be to examine in detail targeted sectors where tort litigation costs are especially high. The pharmaceutical sector would be an obvious candidate, faced with the threat of a large shift of R&D to China and India, in particular, especially to the extent that these countries strengthen the protection of intellectual property for this sector, which is beginning to happen in response to the interests of national companies. With a better understanding of the cost in terms of international competitiveness, a stronger constituency for reform could be built to limit frivolous lawsuits, excessive financial settlements, and extreme criteria for penalizing risk, all of which deter innovation.

Health care.—The number one concern of many U.S. corporate leaders is rapidly rising health care costs, which puts the United States at a significant international cost disadvantage. The fact is that most health care costs are borne by companies in the United States, while in other countries they are usually paid by the government, and at much lower levels in places such as China and India. There is no quick or easy solution to the problem because this basic dichotomy in health care funding will not change, whereas health care costs will continue to rise faster than almost any other cost of doing business.

There can, however, be targeted changes that significantly moderate the rising cost of health care. One important step would be tort reform for the health care sector, as noted above. Many other initiatives are coming from the private sector, involving greater choice and involvement by the employees in making health care decisions, including the basic decisions about cost related to benefits. One model is the government program for federal employees, where costs are shared roughly 50-50 and a number of alternative programs are offered at varying levels of cost and benefits. More innovative approaches involve a given annual financial benefit that, if not used for health care by the employee, can be carried over to the next year or paid, in part, directly to the employee. The government role should be to encourage such innovation toward more cost-effective health care, and should certainly not be to impede it through restrictive regulations.

Energy.—The rising cost of energy is another significant disadvantage for U.S. export competitiveness. High-priced imported oil could be reduced by at least half if domestic drilling were permitted in a barren corner of Alaska and off the Gulf Coast. Environmentalist arguments against such drilling are not substantiated, and in any event need to be weighed against the economic and national security costs of dependence on politically unstable and in some cases unfriendly oil exporters.

For capital-intensive manufactured exports, electricity is generated largely by natural gas, and natural gas prices are far higher in the United States than abroad because of domestic supply shortages. The policy objectives in this case should be greater incentives for increased domestic gas production, improved port facilities for imported liquefied natural gas, and increased nuclear power for generating electricity. This would include eliminating a number of inhibiting government regulations, and leads to the broader problem of overregulation of U.S. export industries.

Regulatory policies.—This is a wide-ranging category of government policy that can impose unjustified costs on American companies and significantly inhibit technological innovation. The magnitude of the problem is daunting. More than 4,000 new "rules" to revise existing regulations are in the bureaucratic pipeline at any one time, and more than 100 of them are likely to cost businesses $100 million or more per year to implement. The total cost of accumulated government regulations, including the voluminous paperwork involved, was estimated at more than $1.1 trillion in 2006, or 9 percent of GDP. An additional $41 billion was spent to administer and police the regulatory system. Considerable power for adopting new regulations and policing existing regulations is delegated to unelected agencies.[165] The broad policy objective should be to reduce and streamline the huge accumulation of government regulations, which would involve more rigorous cost-benefit analysis, more public involvement in rule making, and a more aggressive review and central decision making process within the executive branch. Independence of such decision making from the regulatory agencies is critical because, as explained by Clyde Crews:

> Agencies face overwhelming incentives to expand their turf by regulating even in the absence of demonstrated need, because the only measure of agencies' productivity—other than growth in their budgets and number of employees—is the number of regulations they produce.[166]

The relationship of new, costly regulations to U.S. international competitiveness is almost entirely ignored, and yet the costs in terms of jobs and production transferred abroad and innovation inhibited at home can be substantial. The "Regulatory Report Card" suggested by Crews contains 16 specific grades, but none of them pertains to the

---

[165] See Clyde Wayne Crews, Jr., *10,000 Commandments: An Annual Snapshot of the Federal Regulatory State* (Competitive Enterprise Institute, 2007), pp. 1-2. This annual Crews' report should be required reading for all members of the executive and legislative branches who are involved in expanding the American regulatory state.

[166] *Ibid*, p. 28.

international impact.[167] An international economic impact statement should be a prominent requirement for new regulatory proposals.

One functional area of regulatory policy needs to be singled out as a sobering example of recent overregulation and its serious adverse impact on the international competitiveness of American companies: corporate accounting requirements.

Sarbanes-Oxley.—The rash of corporate scandals at Enron, WorldCom, and elsewhere led to the Sarbanes-Oxley Act of 2002, named after its sponsors, which greatly increased the accounting requirements for publicly registered companies in the United States. Implementation of the act has proven to be extremely costly in a number of ways, thus creating a new comparative disadvantage for U.S. versus foreign production. In retrospect, Sarbanes-Oxley is a classic case of massive and misdirected overkill, and it should be fundamentally revised, based on thorough cost-benefit analysis of its initial years of implementation. The scandals were at the top level of management, and that is where the increased regulation of accounting should focus. Some provisions of Sarbanes-Oxley do this, such as requiring audit committee members of the board to be independent, and for them to hire and oversee the auditors. Restrictions on auditing firms providing consulting services to companies that they audit, which can create a conflict of interest, are also justified. Other provisions of Sarbanes-Oxley, however, do not stand up to reasonable cost-benefit scrutiny.

The exorbitant costs of compliance with Section 404 of the act have gotten the most attention. Auditing costs have roughly doubled, and the annual external cost of compliance runs from an average of $8 million per year for large companies to about a half million dollars for smaller companies, with a relatively higher financial burden on the smaller companies. Additional internal costs are estimated at 100,000 man-hours per year for large companies, with one reporting 130 employees working full-time on Section 404.[168] A survey of 60 member companies by the Manufacturers Alliance/MAPI produced a total cost of compliance during the first year of $389 million, which amounted to 5.8 percent of net income.[169] As for the misdirected focus,

---

[167] *Ibid*, p. 27.

[168] *Economist*, December 18, 2004; *Financial Times*, February 9, 2005; and *Wall Street Journal*, March 2 and 3, 2005.

[169] Donald A. Norman, *The Cost of SOX and the Compliance Process* (Manufacturers Alliance, April 2005). The survey covers a wide range of issues associated with compliance. The follow-up survey, *The Cost of SOX and the Compliance Process—Year 3* (Manufacturers Alliance/MAPI, October 2007), found that total audit costs continue to rise, up 4.1 percent from year 2 to year 3.

the large majority of the costs are not related to revealing a "cooking of the books at the top" but extend from top to bottom, or as one beleaguered manager commented, to some remote warehouse clerk in Timbuktu.

Additional downsides of the act include the requirement that CEOs and CFOs personally certify highly detailed quarterly reports on compliance, which can take extensive time better utilized for corporate leadership. Enormous amounts of information about corporate operations are made public that will never be read by a stockholder but which can be of great value to foreign competitors. Some American companies are going private to avoid the costs of Sarbanes-Oxley, which can limit their ability to finance new investment. Some European companies are de-registering in the United States and moving to London or elsewhere, while new foreign listings on the New York Stock Exchange have been down sharply from previous years.

Sarbanes-Oxley is a disheartening example of overregulation that puts American companies at an international competitive disadvantage. Advanced technology companies are probably hurt the most because they have to be quickest on their feet to stay ahead of their foreign competitors, while Sarbanes-Oxley compliance works to slow down decision making on expenditures at all levels, and makes management generally more risk averse. Foreign economic leadership undoubtedly enjoys reading the reports about the new financial and other burdens imposed on American companies by Sarbanes-Oxley, which can make a shift to production in China, India, and elsewhere more attractive and relatively lower cost.

* * *

This is a broad outline of what the U.S. domestic economic agenda should focus on for maintaining technological leadership and restoring export competitiveness. It is a comprehensive agenda and can be decisive for attaining these objectives. Unfortunately, as noted for almost every issue, the international dimension is either ignored or downplayed. The issues are debated, often in a highly partisan way, with many heavily engaged domestic interests playing influential roles. The role of American multinational companies, however, as global leaders in trade, investment, and technological innovation for 60 years, receive relatively small and often unfavorable attention. This goes to the core of the need to develop a sense of national purpose for restoring international economic competitiveness, and requires serious introspection as to where we are going as a nation.

Our Own Worst Enemy

The acutely observant Pogo is best known for gazing into a mirror and concluding, "I have found the enemy and it is us." This is an accurate reflection of how the United States is reacting today to a new, technology-driven world economic order of rapidly increasing international trade, investment, and technology transfer. It is an amazing era of unprecedented growth in human productivity, with the promise of eliminating global poverty, and American companies have been at the forefront in terms of technological innovation, investment in applied new technologies, and information-age corporate management on a global scale. Such American leadership is the crowning glory of two centuries of national development based on a deeply embedded entrepreneurial culture and economic freedom.

And yet this great human achievement, the envy of the rest of the world, is treated with suspicion, disrespect, and often disdain, by much of the American public and its government. Corporate leaders face unrelenting duress about their motives, if not their morals, while their creative and productive contributions to making the United States the most advanced and affluent nation in history are often ignored. The media accentuate the negative, dwelling on the few bad apples on trial for breaking the law, while Hollywood prides itself in portraying corporate leaders as villains. Public attacks on business leaders are nothing new, and some of the least-productive members of society tend to be the most outspoken critics of the most productive, but such negative public attitudes appear to be on the rise and can now be especially harmful in view of the immediate challenge to U.S. international competitiveness.

Corporate leaders themselves often appear defensive, emphasizing their contributions to charitable causes, in seeming penance for their profits. They should rather state loudly and clearly that it is only through risk-taking entrepreneurship, with many unfortunate losers, that the successful firms have come to market lower cost, attractive new products and services, and that the resulting profits are thus well deserved. Even Bill Gates, the superachiever of the information age, can speak more passionately about his charitable foundation than about the fact that it exists only as the result of his creative, global business leadership in the world of software and information technology services.

Other nations see it differently and pursue forceful economic strategies to develop export-competitive industries, especially advanced technology industries. Priority is placed on technology-oriented education and financial incentives for investment, especially for R&D. China remains a communist state, but successful private corporate leaders, whether foreign or Chinese, and especially those who undertake R&D and produce more advanced technology products, are viewed with admiration and respect. Job-creating entrepreneurship in India is even more welcome. In both countries, American corporate leaders often receive the greatest admiration because of their leading-edge technology performance and their greater respect for labor and environmental standards.

The United States, in contrast, has no clear international economic strategy, and Americans tend to blame America, and its business leaders in particular, for its declining international competitive performance. Other nations may manipulate their currencies, maintain unfairly high import barriers, lower their corporate taxes, and restrain tort lawyers from pursuing frivolous lawsuits against business. But for many Americans, the villains are the American companies that "outsource" American jobs, for the suspect motive of maximizing their profits. Benedict Arnold CEOs!

Two final observations help explain why this is happening and point to a more hopeful turn of events ahead. The first observation is that it has always been part of the American culture—and more broadly that of the English speaking peoples—to be self-critical. A recent history of the English-speaking people during the turbulent 20th century concluded: "Their only possible limiting factor seems to have been a recurring inexplicable, undeserved form of anguished introspection that makes them doubt their own abilities and moral worth."[170]

The second observation is that such anguished introspection is not all bad when coupled with another attribute of the American and other English-speaking peoples, namely that they have the ingrained habit of acting boldly and with foresight once a problem has been sufficiently debated and understood. They are the outstanding problem-solvers by far. The purpose of this study, up to this point, has been, in fact, to help understand the international advanced technology challenge facing the United States today and over the next several years. It concludes with a longer term projection of where the world order of nations is heading, and the vital U.S. leadership role within it, as foresight for meeting this challenge.

---

[170] See Andrew Roberts, *A History of the English-Speaking Peoples Since 1900* (Harper-Collins, 2007), p. 646. The book is a follow-on to Winston Churchill's four volume history through 1900.

CHAPTER 11

THE NEW ASIA-PACIFIC TRIANGLE IN HISTORICAL PERSPECTIVE

The U.S. policy response up to this point has been in terms of immediate issues to be addressed, with objectives stretching out two to five years. These short- to medium-term objectives, however, need to be based on a longer term view as to where the global order of nations is or should be heading. That is the final task undertaken here, and it is indeed formidable in view of the various major historical developments unfolding on highly uncertain paths: the rise of radical Islam and the war on terror; the threatened proliferation of weapons of mass destruction; potential instability in the energy sector and its impact on the environment; a reawakening of authoritarian nationalism; and the principal theme of this study, the emergence of China and India as advanced technology superstates in a global economy being transformed through technological innovation and its wide-ranging application across borders.

No attempt is made at a comprehensive projection, or paradigm, as to where the global political and economic orders are headed. The timeframe is also limited to 10 to 20 years, or to about 2025, since projections beyond that point become exceedingly speculative in critical respects. Important if not decisive historical developments are likely in this timeframe, however, and policy formulation should take full account of them. The presentation, in fact, begins with a discussion of why the period since the end of the Cold War has become and continues to be a genuine transitional period in historical terms. This is followed by a more detailed discussion, emanating from this study, of the projected rise of four regional advanced technology hegemonies that will come to dominate the global economy, although with uncertain consequences for the broader course of history to 2025 and beyond. The final section then addresses the U.S. leadership role as it will or should evolve in the new historical setting.

A Genuine Historical Transition

A transition has been defined as the period between two transitions, but this facile quip does not fit the recent course of human history. There was a sustained historical period of several centuries dominated by the grouping of Western nation states experiencing industrialization, democratization, and imperial expansion, even while almost constantly at war with each other. This period ended with the Second World War when, after a very brief transition, four decades of a bipolar Cold War played out, with the United States and the Soviet Union as the two superpower centers of the NATO and Comecon alliances, pitting Western values of democracy and capitalism against Soviet communism, while many of the weaker, nonaligned nations suffered war and civil strife as the two superpowers struggled for control or alignment. It was, nevertheless, a reasonably stable and well-defined period of history, resulting, in large part, from policies of restraint, or "détente," by the two superpowers. Most observers expected it to continue for decades more.

The sudden collapse of the Soviet Union in 1991 and the peaceful transition of the Soviet bloc toward market-oriented democracy took the experts by surprise and created an intellectual vacuum for defining the new world order of nations. The United States and NATO had clearly triumphed over the Soviet "evil empire," but what would take the place of the Cold War bipolar world order was not clear. The new situation was initially left undefined by the meaningless term "post-Cold War world." In fact, a genuine historical transition had begun and continues to unfold.

After a few years, two competing explanatory paradigms for the new historical course rose to prominence, which were radically different in content and became the subject of considerable debate.[171] The first was the "End of History" paradigm as formulated by Francis Fukuyama, which posited "an unabashed victory of economic and political liberalism" and asserted that "we may be witnessing . . . the end of history as such: that is, the end point of mankind's ideological evolution and the universalization of Western liberal democracy, as the final form of human government." It was a highly optimistic view of global political harmony, and a clarion call for Western triumphalism.[172] The second paradigm painted a far more threatening picture for Western democracies. The "Clash of Civilizations" paradigm, as presented by Samuel Huntington, was based on the hypothesis that "the fundamental source of conflict in the new world will not be primarily ideological or primarily economic. . . . Conflict between civilizations will be the latest phase in the evolution of conflict in the modern world." Western civilization was entering a period of decline and was

---

[171] This synopsis of the two paradigms and how they have evolved and interacted is taken from a broader presentation contained in Preeg, *op. cit.*, 2005, pp. 209-217.

[172] Fukuyama first presented his views in "The End of History?," *National Interest* (Summer 1999), and later elaborated on them at book-length in *The End of History and the Last Man* (Avon Books, 1992). The quotes here are from the article, pp. 1 and 6.

threatened, in particular, from two directions: by the Islamic Resurgence and by the Asian Affirmation.[173]

The threat of Islamic Resurgence has, of course, become a reality in terms of deepening violence within the Muslim world and terrorist attacks against the West, with the United States the principal target, most notably on September 11, 2001, followed by the U.S.-led invasions of Afghanistan and Iraq. This subject is addressed in the following section, particularly regarding the divide between Muslim populations within the four advanced technology regions and across the "Muslim Crescent."

The threat of Asian Affirmation requires greater elaboration in view of its evolution since the early 1990s, and its increasing interaction with the Fukuyama paradigm of economic and political liberalism. The Asian Affirmation was based on a distinctive cultural and political model, stemming largely from Confucianism, that had generated exceptionally high economic growth, initially in Japan and then spreading throughout East Asia, including China. It was a triumphal affirmation of the superiority of the Asian way over the Western model and over American culture in particular. The "Singaporean cultural offensive" contrasted the Asian virtues of order, discipline, family responsibility, hard work, collectivism, and abstemiousness with the self-indulgence, sloth, individualism, crime, inferior education, and disrespect for authority of the West. Singaporean Prime Minister Lee Kuan Yew explained that "the values that East Asian culture uphold, such as the primacy of group interests over individual interests, support the total group effort necessary to develop rapidly." Malaysian Prime Minister Mahathir bin Mohamad described the Asian philosophy more pointedly: "The group and the country are more important than the individual."[174]

This Asian economic triumphalism, however, suffered a setback from Japanese economic stagnation during the 1990s and the Asian financial crises late in the decade. The Asian values of order, discipline, family ties, collectivism, and group over individual rights devolved into crony capitalism, revealing massive corruption and disastrous economic management, particularly in the banking sector, which resulted in the ruin of many companies and extensive job loss. Major economic reforms were undertaken, with great success, in

---

[173] Huntington also presented his views first in an article, "The Clash of Civilizations?" in *Foreign Affairs*, Summer 1993, and later at book-length in *The Clash of Civilizations and Remaking of the World Order* (Simon & Schuster, 1996). The quote is from the article, p. 108, whereas the concepts of Islamic Resurgence and Asian Affirmation were developed in detail in the book.

[174] Huntington, *op. cit.*, 1996, pp. 110, 112, and 114.

restoring economic growth, but they were all in the direction of open markets, more effective competition, the rule of law, and government transparency, or, in other words, the Western economic model. This economic change, moreover, was accompanied by political change, again in the direction of Western values, toward democratic governance, most dramatically in the transition of South Korea and Taiwan from authoritarian regimes to multiparty democracies, with basic individual rights under the rule of law. A distinctive Asian Affirmation thus continues to demonstrate high economic growth and industrial modernization, but now with a mutually reinforcing blend of Asian and Western values.

The most important development within Asia since the early 1990s is the rise of China, and more recently India, toward becoming advanced technology superstates, as elaborated in this study. Neither of the earlier paradigms addressed this issue since Chinese advanced technology industry did not begin to take definitive shape until the late 1990s, followed even more recently by India. Huntington, with reference to China, observed: "The diffusion of technology and economic development of non-Western societies in the second half of the 20th century . . . will be a slow process, but by the middle of the 21st century, if not before, the distribution of economic product and manufacturing output among the leading civilizations is likely to resemble that of 1800."[175] Instead, the rapid process of Chinese industrial modernization has led China to become, by the beginning of the 21st century, the number one exporter of manufactures, predominantly in high-technology industries, with little resemblance to the antiquated trading patterns of 1800.

The historical transition since 1991 has thus been far more complex and dynamic than it appeared in the initial two paradigms, with increasing interaction between them. Other developments of historical importance have also emerged. The proliferation of weapons of mass destruction, with potential access by rogue states and terrorist groups, is now a major threat. India and Pakistan have become overt nuclear powers, with North Korea and Iran close behind. Chemical and biological weapons could become even more threatening. Unstable energy markets and related environmental threats have also broadened and deepened in content, particularly through the rise of China and India as principal consumers of oil and coal, with the most highly polluted cities.

Another troubling development is the reawakening of authoritarian nationalism. National pride and identity can have a positive effect in reinforcing cultural values and can be fully compatible with democratic

---

[175] Huntington, *op. cit.*, 1996, p. 88.

government and individual rights. In more extreme form, however, it can become a justification for authoritarian rule, as happened in the 20th century under the guises of "corporatism" and "national socialism." Such nationalism, with strong xenophobic trappings, can be the last refuge of authoritarian ideologues who have lost their ideology, and this political orientation has been strengthening in China and Russia in particular.

The broadest change of historical consequence since the early 1990s, however, has been the radical transformation of the global economy through the rapid development of new technologies and their transfer across borders through trade and investment. This economic "globalization" is having the most far-reaching historical impact in China and India, as described in this study. Huntington concluded that the fundamental source of conflict in this new world will not be primarily economic. The new technology-driven economic transformation may not be, and indeed should not be, a major source of conflict, but it is rising to become the central driving force for shaping the new world political as well as economic orders.

This, in brief, is the scope of the historical transition under way since the 1990s, which will likely experience further major changes over the coming 20 years. Two decisive dimensions for the outcome, however, will undoubtedly be the rise of what is termed here the four advanced technology regional hegemonies and the evolving U.S. leadership role in global affairs.

Four Advanced Technology Regional Hegemonies

The global economy is increasingly driven by a deepening concentration in and around four advanced technology superpowers: the United States, China, India, and the EU. The first three are also superstates, while the EU, although most of the way to economic and monetary union, is not a "state" in terms of having a unified foreign policy and military. The issue addressed here is how the four will evolve over the coming 10 to 20 years and how this will impact on the overall world economic and political orders.

One fairly certain projection is that each of the four will be a progressively stronger regional economic hegemon. This involves being the dominant regional economy in terms of GDP, trade, investment, R&D, engineers and other technology-oriented professionals, and technological innovation. For others in each region, the hegemon becomes the principal trade and investment partner. The scope of the hegemonic regions requires some judgment, although it is for the most part clearly delineated. The scope projected here has the

United States as regional hegemon for North America and the Caribbean Basin, the EU for Western and Central Europe, including the Ukraine, and Turkey, India for Sri Lanka, Nepal, Bhutan, and Bangladesh, and China for all of East Asia down through ASEAN and Australia/New Zealand.

At this stage, India is far smaller than the other three hegemons, but its relative size will increase in parallel with its much higher national growth, which will likely exceed that of China during the coming 20 years. Looking ahead, three other characteristics of these regional relationships will be determining. The first is that three of the four hegemons—the United States, the EU, and India—are overwhelmingly dominant within their regions, while in East Asia China is clearly number one, but with a more balanced relationship with its regional economic partners. The longer term projection is that the Chinese number one position in the region will strengthen substantially as a result of its size and relatively high rate of growth, as well as the inevitable revaluation of the yuan, but others in the region, particularly Japan and South Korea, will still be in an important counterbalancing position.

The second characteristic is the progressive integration of the East and South Asian regions toward a single Asian region, with two subregional hegemons, China and India. The implications of this integration were discussed in Chapter 6. The projection here is that in 20 years such integration will be well advanced, particularly as a result of the large size and high growth of India.

The third and most important determining characteristic is that over the coming 20 years China and India will almost certainly rise to become fully engaged advanced technology superstates as defined in this study. Per capita income in China and India will remain much lower than that in the United States and the EU, but in terms of advanced technology development and its impact on global economic and military power, China and India will join the United States to become the three advanced technology superstates within a fundamentally changed global political and economic power structure. Among the three, it is unclear whether China or the United States will be the number one economic power in 20 years, and the United States will likely remain number one militarily. But all three will be in a category of their own as advanced technology superstates, far beyond all other nations, unless the EU unifies politically to become superstate number four. The prospect for EU political unification appears dim, with the momentum in recent years in the direction of broadening, at the expense of deepening, the integration process. This could change, however, if a new sense of European identity should emerge from a world increasingly dominated by non-European advanced technology superstates.

The quantitative dimensions of these four advanced technology regions within the global economy are presented in Table 11-1 for 2006. Together they accounted for 71 percent of the global population, 82 percent of global GDP on a PPP basis, and 83 percent of global exports. The 20 year projection is that the population share will remain at about 70 percent, while the shares of GDP and exports will rise toward 90 percent. These increases will result from higher projected rates of growth throughout all four regions, and particularly in Asia. Table 11-1 also shows the shares of global military expenditures and R&D, at 89 and 95 percent, respectively, for the four regions together, while the four hegemons alone accounted for 76 and 72 percent of the global totals. These already extremely high percentages are also likely to rise further over the coming 10 to 20 years.

The four-region concentration of economic size and international engagement, spilling over to military capability as a result of the progressive integration of advanced technology civilian and defense industries, is thus overwhelming in terms of global power and influence. How events will play out within this new order of regional power relationships, however, is highly uncertain. Much will depend on the leadership and policies pursued by the four advanced technology hegemons and, in particular, on whether they act in concert or drift into conflict with one another. The presentation here is limited to comment on the likely directions of change for the most important international relationships.

The Evolving International Economic System

This has been the subject of the preceding chapters, with a policy response framed in terms of two to five years. The short- to medium-term proposals, however, are fully consistent with the longer term evolution of the global economy toward a configuration centered on the four advanced technology regional hegemonies. The objectives of market-based exchange rates and multilateral free trade beginning with the nonagricultural sector would form the foundation for a balanced and open economic relationship among the four regions. This could be decisive in avoiding a separation into protectionist, adversarial regional economic blocs. The content of concerted policies in the areas of international investment, economic aid to the poorest countries, and energy and the environment is less clear, and probably needs at least another five years of further gestation, although the specific issues that need to be addressed have been identified and highlighted.

Table 11-1

Regional Groupings of Countries:
Population, GDP, Exports, Military Expenditures, and R&D Expenditures
(2006)

	Population		GDP*		Exports (Goods and Services)		Military Expenditures**		Research & Development***	
	Millions	%	$ billions	%	$ billions	%	$ billions	%	$ billions	%
North America/ Caribbean Basin	**501**	**7.8**	**16,122**	**26.3**	**1,961**	**21.5**	**640**	**53.3**	**355**	**35.0**
United States	300	4.7	13,195	21.7	1,258	13.8	623	51.9	329	32.5
Europe	**662**	**10.3**	**15,848**	**25.9**	**2,412********	**26.0**	**238**	**19.8**	**246**	**24.3**
EU (27)	493	7.7	13,889	22.8	1,858****	20.4	207	17.2	224	22.1
East Asia	**2,112**	**33.0**	**15,373**	**25.1**	**3,043**	**33.4**	**169**	**14.1**	**323**	**31.9**
China	1,314	20.5	6,286	10.3	836	9.2	65	5.4	136	13.5
Japan	128	2.0	3,628	6.0	703	7.7	42	3.5	128	12.6
South Asia	**1,301**	**20.3**	**2,915**	**4.8**	**170**	**1.9**	**21**	**1.7**	**39**	**3.8**
India	1,113	17.4	2,626	4.3	151	1.7	19	1.6	39	3.8
Four Region Subtotal	**4,576**	**71.4**	**50,258**	**82.1**	**7,586**	**82.8**	**1,068**	**88.9**	**963**	**95.0**
Islamic Crescent	**572**	**8.9**	**3,690**	**6.1**	**711**	**7.8**	**50**	**4.2**	**Neg**	**Neg**
Russia	**143**	**2.2**	**2,193**	**3.6**	**268**	**2.9**	**50**	**4.2**	**22**	**2.2**
South America	**377**	**5.9**	**3,688**	**6.1**	**353**	**3.9**	**23**	**1.9**	**29**	**2.9**
Brazil	187	2.9	1,918	3.1	133	1.5	10	0.8	25	2.5
Sub-Saharan Africa	**760**	**11.9**	**1,366**	**2.2**	**211**	**2.3**	**10**	**0.9**	**Neg**	**Neg**
Global Total	**6,428**	**100.0**	**61,195**	**100.0**	**9,129**	**100.0**	**1,201**	**100.0**	**1,014**	**100.0**

*PPP measure, based on ICP-2005.
**Exchange rate measure.
**PPP measure, based on OECD and other pre-ICP 2005 estimates.
****Extra-EU (27).
Sources: IMF World Economic Outlook Database April 2007; WTO International Trade Statistics 2006; OECD Factbook 2006; Globalsecurity.org for military expenditures; *R&D Magazine*, September 2006, "Global R&D Report 2007"; and 2005 ICP Preliminary Global Report.

The central and decisive issue for the future course of the international economic system is the leadership roles of the four hegemons, together with a few other industrialized economies, including Japan, South Korea, Malaysia, Canada, Mexico, and Brazil. The existing WTO and IMF institutional frameworks, as described in the annexes to Chapters 8 and 9, are at a dysfunctional impasse. At the core of the impasse is the relationship among the four advanced technology hegemons. If they can agree on the existing or revised rules of the game for international finance, trade, and investment, the multilateral system can be resuscitated and strengthened. If they cannot agree, the system will weaken further and likely fragment. This institutional restructuring, moreover, will likely take place over the coming 10 to 20 years, as China and India rise definitively to the status of advanced technology superstates.

A final comment about the longer term outlook for the international economic system is its likely further rise in relative importance within the overall world order of nations, which makes its resurrection and strengthening all the more important. The process of rapid technology-driven economic globalization will almost certainly continue. The extent to which this results in greater relative importance, however, also depends on other events, beginning with the course ahead for radical Islam and the war on terror.

Radical Islam and the War on Terror

This is currently the number one foreign policy challenge for the United States and other Western targets of Islamic terrorism. It is also a terrible struggle within the Muslim world, with Muslims killing Muslims in large numbers, in the context of deep religious conflict and intolerance. The internal Muslim struggle is also political as well as religious in content, pitting secular against theocratic rule in a number of countries. Most deeply, the struggle reflects the frustration of a Muslim culture that was at the forefront of learning and science 500 years ago but has now fallen far behind the West and the Asian East in terms of economic and social modernization in broadest terms.

It is not clear where radical Islam and its related violence will go over the coming 20 years. The level of confrontation and violence has been rising steadily over the past 30 years, although a point may be reached at which the decisive majority of Muslims concludes that enough is enough and the killing must stop. It is encouraging in this regard that from 2002 to 2007 the percentage of Muslim populations supporting suicide bombings has dropped sharply in most Muslim countries.[176]

The commentary here is limited to one geographic relationship within the Muslim world, directly related to the four advanced technology regional concentrations of the global economy, that may have a significant positive effect in turning the tide of radical Muslim hatred and violence. Table 11-1 breaks out the "Islamic Crescent," defined as running from Pakistan and Afghanistan through the Muslim states of Central Asia and the Middle East and across the North African coast to Morocco. This is where the dominant struggle within the Muslim culture is taking place, and where the training ground for terrorism is concentrated. In 2006, the Muslim Crescent accounted for 8.9 percent of the global population but only 6.1 percent of the global GDP, despite the high price of oil which is the principal source of wealth for a number of these countries. This relatively low share of global GDP is a measure of the relative economic decline of the Muslim world and it will almost certainly decline further relative to the four advanced technology regions as long as the religious struggle and violence continue to deter productive, job-creating investment and economic modernization. The decline would be even faster if the price of oil were to drop sharply, which is likely to happen at some point over the coming 10 to 20 years.

The other noteworthy and related observation is that the 572 million population within the Islamic Crescent constitutes roughly half of the global Muslim population, with almost all of the other half within the four advanced technology regions: in Turkey within the European region, in Malaysia and Indonesia, among others, in East Asia, and in Bangladesh and India within South Asia. What this means, over the coming 20 years, is that there will be a clearer and clearer divide between the affluence, higher education, and growing middle-class ranks of the half of the Muslim population that lives in the advanced technology regions and the slower pace of modernization, if not stagnation, for most of the Islamic Crescent half of the Muslim population. The question is whether the half in the advanced technology region will have a positive influence, by demonstration effect, on the more troubled half within the Islamic Crescent, moving the latter in the direction of secular democratic government, religious tolerance, and economic modernization. The observation here is that

---

[176] See the Pew Global Attitudes Project, *Global Opinion Trends 2002-2007: A Rising Tide Lifts Mood in the Developing World* (July 2007) at www.pewglobal.org. From 2002 to 2007, the percentage of Muslims viewing suicide bombings as justified dropped from 74 percent to 34 percent in Lebanon, from 44 percent to 20 percent in Bangladesh, from 33 percent to 9 percent in Pakistan, and from 26 percent to 10 percent in Indonesia (p. 55).

there are grounds for optimism that this positive movement will happen, at least to some degree.

Moreover, the pivotal nation in this Muslim divide is India. India has 150 million Muslims who continue to suffer discrimination in many ways and who lag greatly in education, largely of their own making because of fundamentalist religious schools. But now, on the path of 8 to 10 percent annual growth, driven by a private sector that is creating jobs where there is far less discrimination against Muslims, a more affluent and democratized Muslim middle class is emerging and should grow substantially over the period projected here. This process of change can be illustrated by an extraordinary contrast. Azim Premji, the founder of the multinational business services company Wipro, is a Muslim entrepreneur. Trained as an engineer, he transformed the family vegetable oil business into a multibillion dollar advanced technology company, and became a billionaire in the process. Mohamed Javeed, principal of Bangalore's predominantly Muslim Al-Ameen College, refers to Premji as an icon and role model. And yet, only a few miles from Premji's Bangalore home, students in the Masjid e Takwa madrassa spend their days memorizing the Koran in Arabic, a language that neither the students nor the teachers understand, while courses in neither science nor English are offered.[177] These unfortunate students are being condemned to a life of continuing poverty, but Muslim parents throughout India have to see that there is now a far better alternative for their children, with Azim Premji the outstanding role model.

The policy response to this particular dimension of radical Islam and the war on terror clearly lies in the realm of "soft power," and consists of communicating to Muslims on both sides of the advanced technology regional divide that education and economic modernization, stemming from the rule of law and religious tolerance, is the preferable way to go. As for this potentially critical direction of change, the lead policy role, by demonstration effect at home and by outreach in communicating with neighboring Muslim countries, can best be played by India.

Other Regions

The other regions listed in Table 11-1—Russia, South America, and Sub-Saharan Africa—account for 20 percent of the global population but only 12 percent of GDP and exports, and 7 and 5 percent, respectively, of military expenditures and R&D. They should

[177] See Yaroslav Trofimov, "How a Muslim Billionaire Thrives in Hindu India," *Wall Street Journal*, September 11, 2007.

all benefit from continued rapid growth in trade and investment generated by the four advanced technology regions. They will likely drift further behind in relative terms, however, because of shortcomings in policy frameworks and other factors for generating high levels of job-creating investment. Each one also has distinctive characteristics in its relationships with the four advanced technology regions.

Russia.—Russia currently benefits greatly from high oil and natural gas prices, and remains important in international affairs as a nuclear power and a permanent member of the UN Security Council. Otherwise, however, Russia faces major downsides that are likely to become more serious over the coming 20 years. It has a demographic problem, with an overall population in absolute decline together with a large and growing Muslim minority, which is likely to cause ethnic tensions. The industrial sector is largely obsolete and noncompetitive, while serious environmental problems need to be addressed. The political trend toward authoritarian nationalism breeds corruption and crony capitalism, while economic reforms languish. These various problems need to be addressed because together they could lead to a serious economic setback, especially if oil prices decline. The basic choice for Russia is to move definitively to Western political and economic liberalism and deepening integration with the EU or to maintain an independent, more authoritarian nationalist course, sandwiched between the EU and Chinese advanced technology superpowers. The former course offers clear advantages, but the latter, nationalist course, with deep historical roots, could well prevail.

South America.—Development experts have mixed opinions as to why South America in general, and Brazil in particular, have been unable to achieve sustained high levels of job-creating investment and growth. The lack of structural economic reforms and the very wide distribution of income related to tax evasion and corruption comprise much of the problem. Political swings back and forth from left to right prevent sustained economic reform strategies from being implemented, as they are in China and India. Looking ahead, a defining characteristic will likely be the growth path of Mexico and the Caribbean Basin, more and more deeply integrated with the United States through free trade and investment, compared with that of South America. Structural economic reforms and improved law enforcement leading to higher economic growth appear more likely in the former, which could increase pressures in South America to follow suit. A particularly interesting development will be the post-Castro course in Cuba. The projection here is that Cuba will move relatively quickly to democratization and free trade with the United States, with the result of becoming, to a large extent, a subregional extension of the Florida peninsula.[178] This, in turn, would have a catalytic effect for accelerating the overall integration of the North America/Caribbean region.

Sub-Saharan Africa.—The development challenge in Africa centers on the need for structural economic reforms and good governance together with elimination of widespread political violence. Economic aid from the four advanced technology regions should be plentiful to the extent that it can be absorbed effectively, and foreign investment is rising, including from China and India, particularly for oil and other natural resource development. In recent years, a few African nations have achieved sustained growth and better governance, but the 20-year outlook for the region is unclear.

Nonproliferation, Energy, and the Environment

These are all areas of great international importance, but no attempt is made here to project their 20-year course ahead. Discussion of the short- and medium-term outlook was provided in earlier chapters, particularly for the relationship among the four advanced technology superpowers. The point of departure was that there are broadly-based common interests for curtailing the proliferation of weapons of mass destruction, achieving greater energy security at lower prices, and improving environmental standards. Nevertheless, in each case, the prospect for concerted actions to these ends was found to be relatively weak over the short to medium term. This could change over the longer, 20-year time frame, however, toward more closely coordinated actions. These responses would likely focus heavily on joint development and application of new technological capabilities, in terms of surveillance for weapons of mass destruction, lower-cost cleaner sources of energy, and less energy-intensive consumption.

Leadership In Concert or In Conflict

The degree to which the four advanced technology superpowers exercise concerted leadership, or drift into conflict, will almost certainly be the most important factor in shaping the global political and economic orders over the coming 20 years and beyond. Interests will vary on individual issues, but the existing broad base of common interest should strengthen over time as China and India progressively rise to become more affluent, advanced technology superstates, with a large middle class at the center of the national economic and political

---

[178] The forces in play are described in Ernest H. Preeg, *Cuba and the New Caribbean Economic Order* (CSIS, 1994).

power structures. How this can translate into more concerted international leadership, however, will depend, perhaps decisively, on bridging what can be referred to as the two great divides: the North/South divide in economic terms and the democracy/authoritarian divide in political terms. The heroic projection made here is that, over the coming 20 years, developments will favor the bridging of both of these great divides.

The North/South economic divide.—This issue of the dichotomy between developed and developing countries has been addressed in detail for current difficulties within the international financial and trading systems. Among the four advanced technology superpowers, it pits the United States and the EU against China and India. The projection here is that this long-standing divide will progressively weaken among the four over the coming 20 years in those areas of international economic policy where it is most heavily concentrated. The United States and the EU will continue to have much higher per capita incomes, even after very large revaluations of the yuan and the rupee, but in terms of economic modernization and power, including trade competitiveness and international investment, relationships will become far more balanced, with China and India, indeed, stronger than the United States and the EU in some areas.

The fading away of the North/South economic divide among the four is already happening to some extent. Economic aid, which was a major issue of transferring wealth from the rich countries to the poor, is losing significance for China and India as aid recipients, and they are both likely to shift officially to the donor side of the table over the coming 20 years. The North/South divide in the more critical areas of trade and financial policies, despite current difficulties in the Doha Round and over currency manipulation, are also likely, over time, and as proposed in this study, to move in the direction of multilateral free trade and market-based exchange rates, most importantly among the four advanced technology regions, which would effectively end the North/South divide in these policy areas. Concerted action to improve the environment, including a reduction of CO_2 emissions, appears the most intractable issue in the short run but again, as noted above, the longer term solution will likely center on the development and application of new technologies, which can best be achieved through concerted action among all four advanced technology regions, without roadblocks related to being more or less developed.

The democracy/authoritarian political divide.—This is the more threatening divide, between the three democratic economic superpowers and authoritarian China. The outcome will also be far more important. The current divide, with China rising to become the number two global military power in a largely adversarial relationship with the

United States, constitutes not only a major barrier to collaborative actions in a number of areas, but also a risk of armed conflict over Taiwan. The anticipated rivalry among the U.S., Indian, and Chinese blue water fleets is a striking harbinger of what lies ahead in broader terms if the political divide continues.

The outlook for Chinese democratization has an internal and an external dimension. The more important internal dimension involves the rapid expansion of an educated, more affluent middle class and the increasing difficulties of managing a highly complex advanced technology economy without the independent rule of law and other checks and balances on self-serving government actions and corruption. The suppression of individual freedom becomes less and less defensible and, most fundamentally, authoritarian rule by nine self-selected, aging ideologues who have lost their ideology[179] loses its justification.

The external dimension will play out to a large extent within Asia. The deepening of democratic roots throughout the region, from Japan and South Korea through ASEAN to India sharpens the contrast with Chinese authoritarian rule. Democratic governance in Hong Kong and Taiwan likewise brings the democratic path within China itself.

The United States will also play an important role in the hoped-for peaceful democratization of China. The bilateral relationship between what will soon be the two global superpowers is already structured, in organizational terms, up through the top levels of government and needs to be managed so as to address, in a supportive way, the Chinese government's stated objective of democratization. This will not be a simple task, and leads to the broader question of what the U.S. leadership role should be in the emerging new world of technology-driven economic transformation and advanced technology superstates.

The American Leadership Role

The U.S. leadership role is a fitting subject on which to conclude this study. American global leadership has been decisive over the past century in defeating three assaults on Western values of democracy, individual rights, and economic freedom, in two world wars and one Cold War. The United States is currently playing the lead role in a fourth assault from radical Islam and international terrorism, a defense that is deeply troubled and subject to strong criticism at home and from abroad. Domestic sentiment to turn inward and external efforts to reduce U.S. power and influence have been growing.

---

[179] A reference to the Politburo Executive Committee.

The basic conclusion here is that a substantial decline in the U.S. international leadership role would be a great mistake for all. The need for international leadership is greater than ever before in a globalizing world power structure that is being transformed by amazing new technologies, which results in potential great material benefits together with a Pandora's box of grave security threats. The form and substance of U.S. leadership, however, will need to change and adjust to the new realities.

The period of the United States being the single superpower is proving to be relatively brief, indeed one part of the transition between the bipolar Cold War world and some new multipolar power structure, albeit with the United States still primus inter pares in key respects. The transformation of the power structure, as argued here, centers on the rise of regional concentrations in advanced technology development and application, with impact on military as well as economic power relationships. Table 11-1 encapsulates the story in starkest terms. This transformation also has an historical geographic dimension, with the long-standing North Atlantic dominance by the United States and West Europe displaced by a three-way relationship that includes Asia.

The exercise of U.S. leadership within this new order of power relationships will continue to rest heavily on the core grouping of industrialized democracies—West Europe, Canada, Japan, South Korea, and Australia. The relationships with China and India will progressively grow in importance, however, and over the coming 20 years will become essential components for decisions on almost all major international issues. U.S. joint leadership with India and China thus becomes a principal challenge for the United States. It is joint leadership among the three advanced technology superstates within the new Asia-Pacific triangle.

The two U.S. bilateral relationships within this triangle are complex, and in terms of concerted leadership they will evolve slowly. The United States and India share basic values with respect to democracy, human rights, and economic freedom, but Indian foreign policy since independence has been driven by a deep commitment to nonalignment, to separation from all major foreign powers. This should progressively change as India becomes more heavily engaged globally in economic terms, which will inevitably spill over into political and security areas. The challenge for the United States is to strengthen collaboration on multilateral interests while respecting the Indian commitment to basic independence from foreign powers.

The U.S.-China outlook is far less clear as to direction and outcome. The bilateral relationship is deeply engaged in almost all areas and has become the single most important U.S. bilateral relationship. The longer term course ahead, however, including the degree of collaborative leadership in global affairs, will be fundamentally different, depending on whether China remains an authoritarian state or crosses the divide to democratic governance. Nothing more is offered here than to state that U.S. encouragement and positive support for the democratic alternative in China, as proposed earlier, should be a persistent top priority for U.S. foreign policy.

There is finally the fact that international leadership is based not just on substantive content and power relationships, but also on how the power is exercised. The proclivity to global leadership in support of international norms and commitments varies greatly among nations. In a few nations the proclivity is deeply embedded by habit and culture, whereas most nations limit their engagement to pursuing more narrowly defined self-interests. Such a global leadership culture can have an heroic quality of taking on responsibilities and costs, including the cost of citizens' lives, to achieve broader international goals.

The United States has clearly exercised such global leadership over most of the past century. Within Europe, the United Kingdom has usually played the most prominent global leadership role, reflecting back to its 19th century assumption of what was then called the "white man's burden." In Asia, India has stood out over 60 years since independence for playing an international leadership role well beyond its relatively weak power position.[180] The Chinese leadership role, in contrast, has been far less directed to global norms and values, as in recent experience in pursuing self-interest in petroleum and raw material supplies at the expense of international objectives in the areas of nonproliferation and basic human rights.

Looking ahead 20 years, with all the new and uncertain global forces in play, it is frightening to contemplate what might happen if the United States drew back to wait for international consensus before acting. This is, of course, a straw-man premise. The United States will continue to play a strong leadership role, but it must adapt to a new order of power relationships and priority objectives. The U.S.-Europe-Japanese foundation of shared security and economic objectives needs to continue, while being extended to include India and China, in particular, with the United States best positioned to bring about this extension through the new Asia-Pacific triangular relationship. The longer term objective for all of the global leaders should be to continue

---

[180] It is not accidental that all three cited global leaders are part of the English-speaking world. Much of India's leadership in the nation's formative years was the product of education in England, and the United States is another former English colony. Looking ahead, a closer relationship among these three habitual global leaders may develop, based on shared values and a common language. This subject is worthy of study by associated English-speaking think tanks.

technology-driven economic growth and modernization and the alleviation of global poverty, to curtail international terrorism and the proliferation of weapons of mass destruction, and to build the habit of joint leadership based on the common values of economic freedom, basic human rights, and democracy. In broadest terms, the objective can thus be phrased as the delayed triumph of the End of History paradigm.

COMMENTARY

INDIA—TRANSFORMED AND ETERNAL

by Teresita C. Schaffer*

India has undergone a transformation in the past decade and a half. Increasingly rapid economic growth—over 9 percent per year for the past two years—has expanded the middle class, created an expectation of material success for the country, and launched new international and political ambitions. The disintegration of the political institution that was the Congress Party and the rise of single-state political parties has laid the groundwork for one coalition government after another, changing the dynamic for managing economic policy. The new political prominence of social and caste groups that had previously been relatively passive "vote banks" has shaken up the power structure and created expectations of wider participation in the gains of growth. The end of the Cold War led to a foreign policy metamorphosis, with the United States emerging as India's most important external friend, China appearing as both rival and partner, and Russia's importance diminishing palpably. India, now a nuclear power as well as an economy on the move, is clearly becoming a power with global significance.

Ernest Preeg's account of the start of India's rise to the status of "advanced technology super-state" describes the economic side of this success story. Like him, I am optimistic that this growth will turn out to be a durable phenomenon. India's rise, which started in the services and especially information technology sector, has spread in recent years to manufacturing. It is based largely on productivity growth. Because of the increasing importance of private investment in India's economy, it has become more self-sustaining and less dependent on the day-to-day evolution of government policy than was once the case. And India's economy has led the way in integrating India with the rest of the world. Thus, India's foreign policy increasingly reflects the country's drive to expand trade, investment and energy sources, and this economic "driver" in turn is producing a more focused and pragmatic Indian approach to the world.

India does not altogether match the definition Preeg gives of "advanced technology superstate." Under his definition, such a state starts by developing scientific and technological excellence, and

* Director, South Asia Program, Center for Strategic and International Studies.

proceeds from there to financial prominence and finally to military power. In India's case, the sequence is different. India has been the primary military power in its part of Asia since it became independent in 1947. Military modernization has been a part of its recent drive, and its economic growth will doubtless make it easier to sustain the hoped-for expansion of military power. And I'm not sure whether India's behavior—a blend of public policy and private economic activity—will really match the purposeful strides suggested by the term "superstate." But clearly, India is already an important regional player and is becoming a global one.

Preeg's account discusses two of the "speed bumps" in the way of India's rise that are often cited: the need to upgrade education and infrastructure. But what kind of international and economic actor India turns out to be will also be affected by two factors that are beyond the scope of this book.

The first is the problem of those left out of India's exciting rise—both socially and geographically. Looking up and down the income scale, there are "three Indias." The "superstate candidate" consists primarily of people and enterprises that form part of the most advanced India, what some observers have called "airplane India." The go-go Information Technology sector, globalized industries, the internationally integrated R&D establishment, the upper and upper middle class families who almost without exception have members who either live in the West or have received part of their education there—these are, collectively, the engine of India's technology-driven growth. They have generated much of the dynamic growth of the past 15 years; they are pushing government policy forward; they are measuring India against a global standard, and doing their best to ensure that it can compete in the big leagues.

Side by side with "airplane India," however, are two others. Middle class India—"scooter India," to continue the transportation analogy—is perhaps 250 million strong. Many of these people have benefited from India's greater prosperity, and have eagerly sought jobs in the emerging tech sector. If India continues its rapid growth, as most observers expect it to at least for the next several years, "scooter India" will provide much of the fuel for that growth. This group's education will increase, its earning power will grow, its eagerness for jobs in manufacturing and in the globalized sector will intensify—and perhaps more importantly, its numbers will rise. This group is already educated at least through the primary level. The challenge will be to make secondary education universal in this group, and to make universities and advanced training institutions increasingly accessible to it.

It is the third India, "bullock cart India," that represents the most serious potential drag on India's rise. Much of this group depends on agriculture for all or most of its livelihood. Agriculture has lagged behind industry and services in this recent time of prosperity. Even taking into account increasing urbanization, agricultural livelihoods have grown more slowly than the national norm. Literacy has increased in the rural areas, but village India has modernized slowly. Prosperity depends on the weather; individuals feel very dependent on patronage and on government programs for benefits that may make life a bit better; and old social relationships die hard.

Even more politically salient are the geographic inequalities that go along with this three-part social division. Per capita income in Maharashtra, which boasts the highest per capita GDP among India's large states, is close to five times the average income in Bihar, the state that consistently appears at the bottom of the charts. In parts of India where growth has been high, villages have been pulled along in its slipstream. But in several of the country's largest states, a pattern of low growth, lagging human development, and weak literacy is intertwined with corrupt and patronage-ridden governance. India can continue its growth for some time despite these inequalities, but at some point village poverty and especially the poverty of some of the country's largest states will drag it down.

India's political leaders are well aware of this, and respond with populist policies, including the Congress-led government's program for guaranteed work income for rural families. India's government needs to meet two difficult challenges in order to continue the country's astonishing economic transformation over the long term. It needs to find an effective way of pushing some of the growth down to the grass roots. It also needs to make its combination of policies attractive enough to be voted back into office. The Indian political system is hard on incumbents. In 2004, the winning Congress party actually lost vote share, and of the 98 members it had had in the outgoing parliament, only half were reelected. This is the policy dilemma that each government in turn will have to confront.

The second big question mark has to do with India's web of international relationships and the changing patterns of political, economic and security relations within Asia. India practiced its own version of "splendid isolation" during the years when it foreign policy was based on nonalignment. Russia was its most important external partner; relations with the United States were for the most part relatively thin and distant; and relations with China went from high rhetorical warmth to hostility after India's defeat in a war with China in 1962. Since the 1990s, the pattern has changed dramatically. The United States has become India's most important external friend; relations with China involve engagement and rapidly growing economic ties, although the security rivalry is still there; and Russia is still a major military supplier but does not have the weight it once did.

As part of this foreign policy transformation, India has also become much more deeply engaged in Asia. Trade has increased across the board. Japan, once primarily an aid donor with little political dialogue, has become an important political relationship. India has cultivated military ties not just with the United States but with a number of East Asian countries—particularly Singapore, but also Vietnam, Japan and Australia. Modest exercises with China reinforce the policy of partnership that both governments are emphasizing. India is beginning to find its way into some of the Asian regional institutions: it is a dialogue partner with ASEAN; it is a member of the East Asian Summit, despite China's efforts to marginalize it. It is not a member of APEC, and the United States has shown little enthusiasm for bringing India in, although in many ways today's economically driven India would be a natural addition to the group.

India is, in short, a major Asian regional power—but neither India nor the region is clear at this point on what "regional Asia" is becoming or how India will participate. India's growing ties with its neighbors to the East build on its ties with the United States, and India's interests in Asia overlap to a significant extent with those of the United States. At the same time, India represents a large potential new entrant into the East Asian "club," one that would complicate existing power relations there. And India may turn out to be ambivalent about a central U.S. objective: keeping the United States intimately involved in Asian regional life and regional institutions.

East Asia represents a region of great promise for Indian policy-makers. The Middle East, by contrast, has great strategic importance and serious potential trouble. India has major economic relationships in this area, including some two-thirds of its oil imports, and is cultivating a strategic presence there as well. India and the United States have significant policy differences, particularly regarding Iran. None of this is likely to affect India's technological and economic expansion, but it will affect the way India relates to the world's major powers and hence the way it participates in the global strategic and economic dialogue.

In both these key regions and more broadly in the global arena, one of the guiding principles of India's foreign policy is "strategic autonomy." Indian policy makers of all political persuasions, and those voters who care about foreign policy, are committed to maintaining the maximum freedom of action for India in the world. As India develops a strategic partnership with the United States, its leaders are also intent on reassuring India's elite audience that they have not given up control of India's foreign policy to anyone, and specifically not to the United

States. This adds intensity to India's quest for positions of global leadership (such the recent election of veteran diplomat Kamalesh Sharma as Commonwealth Secretary General). It complicates the task of having a serious policy dialogue on contentious issues, such as Iran policy. Indians and their close international partners are trying to develop a new model for international partnership. The models honed during the Cold War years will not be suitable for U.S.-India relations.

India's ties with China are ambivalent in different ways. China is India's biggest strategic rival, but is also devoted to the kind of multipolar world that appeals to Indian policy-makers. China is a major trading partner of India, but increasingly, it is starting to run persistent trade surpluses, so that the trade relationship is acquiring an edge of rivalry. China talks about partnership with India, but has cultivated strategic ties with Pakistan going back several decades.

As a result, the U.S. and Indian governments spend much time fending off charges that the United States is trying to enlist India in an international anti-Chinese conspiracy. For the same reason, I am skeptical about Preeg's argument that an India/China/U.S. triangular relationship will become a major motor of global economic and security relations. Clearly, each of the countries in this triad will be watching the other two carefully. But the three-way relationship will have strong overtones of balance of power politics, and will involve shifting ties with larger powers beyond the triad, especially Japan and Russia. Rather than a three-way cooperative nexus, I anticipate that the next couple of decades will bring a more fluid set of economic power relationships that will tax our diplomatic skills. The big prize, if we handle this effectively, will be the possibility that the world can manage the rise of India and China without undue military or political conflict. This would be a unique historical achievement, and is worth devoting considerable intellectual and diplomatic resources to the task.

INDEX

ABOUT THE AUTHOR

Ernest H. Preeg holds a Ph. D. in economics from the New School for Social Research and was a career foreign service officer specializing in international trade, finance, and economic development. He was a member of the U.S. delegations to the Kennedy and Uruguay Rounds of trade negotiations, and served as Deputy Assistant Secretary of State for International Finance and Development, Chief Economist at the U.S. Agency for International Development, and White House Executive Director of the Economic Policy Group. He also served as the American Ambassador to Haiti.

Since retirement, he held the Scholl Chair in International Business at the Center for Strategic and International Studies (CSIS), was Senior Fellow at the Hudson Institute and, since 2000, Senior Fellow in Trade and Productivity at the Manufacturers Alliance/MAPI.

His recent publications include:

- *Traders in a Brave New World: The Uruguay Round and the Future of the International Trading System* (University of Chicago Press, 1995)
- *From Here to Free Trade: Essays in Post-Uruguay Round Trade Strategy* (University of Chicago Press and CSIS, 1998)
- *Feeling Good or Doing Good with Sanctions: Unilateral Economic Sanctions and the U.S. National Interest* (CSIS, 1999)
- *The Trade Deficit, the Dollar, and the U.S. National Interest* (Hudson Institute, 2000)
- *From Here to Free Trade in Manufactures: Why and How* (Manufacturers Alliance/MAPI, 2003)
- *U.S. Manufacturing: The Engine for Growth in a Global Economy*, co-editor with Thomas J. Duesterberg (Praeger Publishers, 2003)
- *The Emerging Chinese Advanced Technology Superstate* (Manufacturers Alliance/MAPI and Hudson Institute, 2005)